Ash from Coal and Biomass Combustion

Ashok K. Singh · Reginald Ebhin Masto ·
Bodhisatwa Hazra · Joan Esterle ·
Pradeep K. Singh

Ash from Coal and Biomass Combustion

Springer

Ashok K. Singh
Council of Scientific and Industrial Research—Central Institute of Mining and Fuel Research
Dhanbad, Jharkhand, India

Bodhisatwa Hazra
Council of Scientific and Industrial Research—Central Institute of Mining and Fuel Research
Dhanbad, Jharkhand, India

Pradeep K. Singh
Council of Scientific and Industrial Research—Central Institute of Mining and Fuel Research
Dhanbad, Jharkhand, India

Reginald Ebhin Masto
Council of Scientific and Industrial Research—Central Institute of Mining and Fuel Research
Dhanbad, Jharkhand, India

Joan Esterle
School of Earth and Environmental Science
University of Queensland
St Lucia, QLD, Australia

ISBN 978-3-030-56980-8 ISBN 978-3-030-56981-5 (eBook)
https://doi.org/10.1007/978-3-030-56981-5

This Springer imprint is published by the registered company Springer Nature Switzerland AG
The registered company address is: Gewerbestrasse 11, 6330 Cham, Switzerland

Coal and Biomass: More than just energy resources

Preface

Generation of huge amounts of ash from combustion of coal and biomass in power plants is a major concern for the environmental managers and technologists of the plants. China, India, USA, Japan, and Korea are the major coal producers of the world. Recent regulations in many countries demand 100% utilization of the ashes produced after combustion for generating energy. The heterogeneous nature of the coal and biomass ashes, wide varieties of the utilization options available, and the environmental issues associated with the ash made the subject of ash very interesting and challenging. Our research on coals and biomass from various perspectives in recent years has led us to realize that although a substantial body of research has built-up over many decades especially on their combustion aspects, a comprehensive discussion with these two resources together in a same platform has been missing. Such realization has stimulated us to compile this monograph.

The full domain of coal and biomass ash is presented in the book, under five chapters. The purpose of the monograph is to give a fundamental understanding of the ash properties, generation, bulk and value-added utilization sectors, and the environmental issues. This piece of work would be a valuable repository for wide spectrum of audience covering students, managers, environmentalists, and policy makers associated with ash and environment. In addition, this compilation of all the information on ash will help the researchers to get a quick overview on the subject matter.

We introduce the subject in Chap. 1 covering the importance of coal-based energy and the science behind the ash generation in power plants. Both coal and biomass generate ash of very distinct nature, volume, and properties. The detailed characterization covering physical, chemical, and other special properties of the ash are detailed in Chap. 2. The different options available for the use of ash along with the basic minimum description of all the utilization sectors and the suitability of ashes for each sector are discussed in Chap. 3. Chapter 4 addresses the environmental issues associated with the fly ash use along with the basic chemistries of the origin of potentially toxic elements in ash, their leachability and potential hazard, and finally the long-term fate of the ash and the different abatement options available to decrease the environmental concerns. The key outcome of this compilation is summarized in Chap. 5.

We hope this monograph would be an interesting contribution in the field of coal and biomass ash sectors, and excites the readers to think more about ash.

Dhanbad, India
June 2020

Pradeep K. Singh

Acknowledgements The Director, CSIR-Central Institute of Mining and Fuel Research, India is acknowledged for granting permission to publish this work, and also for providing necessary infrastructure to carry-out the work.

Contents

Chapter 1
Introduction

Abstract Coal, a naturally occurring laminated organic-sedimentary rock, has been a source of energy for several decades. Being composed of essentially lithified-plant materials, which were initially deposited in swampy environments, coal is a unique fossil fuel used for a variety of industrial and domestic purposes. Likewise, biomass is a renewable solid fuel used for electricity generation and for meeting the domestic and industrial thermal energy needs. Energy extraction by combustion of coal and biomass generates large amount of ash, which is a solid waste. The quantum of ash generated from coal and biomass sector and their potential utilization is discussed in this chapter.

Coal, a naturally occurring laminated organic-sedimentary rock, has been a source of energy for many since several decades. Being composed of essentially lithified-plant materials, which were initially deposited in swampy environments, coal is a unique fossil fuel used for a variety of industrial and domestic purposes. The primary use of coal till date is to produce steam for electricity generation through combustion.

The major coal-producing countries across the world are China, India, USA, Australia, and Indonesia. Coal is mainly used for generation of electricity in power plants. Country-wise coal production during the year 2018 (www.bp.com 2019) is given in Fig. 1.1. In spite of the bloom of renewable energy resources, globally, coal remains as the main source for electricity generation. Worldwide, coal accounts for an average contribution of 37% of the total electricity generation. Ever since from the establishment of first commercial power plant in the year 1822 at Pearl Street Station, New York, USA, the coal-based electricity generation has developed tremendously. Currently, China, USA, India, Japan, and South Korea are the topmost coal-based electricity-generating countries (Table 1.1).

Coal, for some time now, has attracted the attention of several researchers and policy makers due to its negative impact on atmosphere (Ward and Suárez-Ruiz 2008). While the importance of coal in meeting energy requirements is hardly doubted, the rising CO_2 levels in the atmosphere has been a point of global concern. The contribution of coal combustion in CO_2 release to the atmosphere has caught the eyes of the world (World Coal Institute 2005; U.S. Environmental Protection Agency 2005; Commission of the European Communities 2005). Consequently, in

A. K. Singh et al., *Ash from Coal and Biomass Combustion*,
https://doi.org/10.1007/978-3-030-56981-5_1

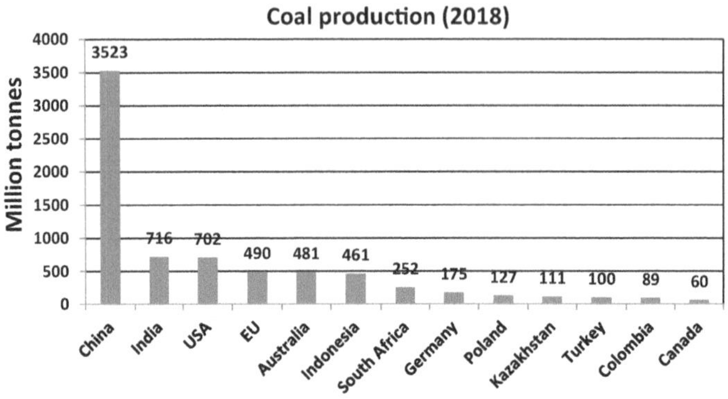

Fig. 1.1 Major coal-producing countries of the world

Table 1.1 Data of the top five countries that use coal for electricity generation (*Source* https://www.iea-coal.org/top-coal-fired-power-generating-countries/

S. No.	Country	Coal fired electricity (TWh)	% of total electricity	Major power stations/plants (PS/PP)
1.	China	4,360.9	66%	6.7 GW Datang Tuoketuo PS 5.1 GW Waigaoqiao PS 5 GW Guodian Beilun PS
2.	USA	1,314.0	30%	3.6 GW Robert W Scherer Power Plant 3.3 GW Monroe Power Plant 3.37 GW Plant Bowen
3.	India	1,141.4	60%	4.7 GW Vindhyachal Thermal PP 4.6 GW Mundra Thermal PS 4 GW Mundra Ultra Mega PP 3 GW Talcher Super Thermal PS
4.	Japan	342.5	32.3%	4.1 GW Hekinan coal PP 1.8 GW Maizuru PS 1.6 GW Tomato-atsuma PS 1.4 GW Reihoku PS
5.	South Korea	264.4	43.1%	6.1 GW Taean PS 6 GW Dangjin Thermal PP 5.08 GW Yeongheung PS

recent years, alternate energy sources and new technologies and amendments have been made with a focus on improving the efficiency of combustion and power generation and concomitantly reducing the CO_2 production per unit of electrical energy produced (Edenhofer et al. 2011). To combat greenhouse gas emission, carbon neutral approaches like biomass combustion and biomass co-firing are practiced in many countries. Furthermore, it has also been opined that adopting and deploying carbon-negative approaches (viz., application of bioenergy with carbon capture and storage

Table 1.2 Country-wise electricity generated from biomass and waste (*Source* IRENA 2018)

	Country	Year	Biomass power	
			GWh	% of total
1	USA	2016	69,017	1.6
2	Brazil	2016	51,040	8.8
3	China	2016	49,403	0.8
4	Germany	2019	44,420	8.6
5	UK	2018	34,759	10.4
6	Italy	2016	19,509	6.7
7	Thailand	2016	17,672	9.2
8	Japan	2016	16,847	1.6
9	India	2018	16,325	1.1
10	Canada	2016	12,385	1.9
11	Finland	2016	11,523	16.8
12	Sweden	2016	11,487	7.4

possibilities) is likely to play a significant role in the second half of this century (Clarke et al. 2014; Masson-Delmotte 2018; Lu et al. 2019).

Unlike coal, biomass is a renewable source of energy. Biomass is plant or animal material used for production of electricity or heat. Many countries are at the transition to biomass-based power to meet the sustainability goals to provide electricity to their industries and household. There are varieties of biomass feedstock with diverse properties. Some of the biomass feedstocks are forest residues and wood waste; agricultural residues (corn stovers, wheat stalks, etc.); urban wood waste (packing crates, pallets, etc.); municipal solid waste; food processing residues; and so on (Irena 2012). The global contribution from biomass fired electricity generation is 130 GW. Supply of biomass is highest in Asia, followed by Africa and American continents. Countries that contribute significantly to biomass-based electricity generation are listed in Table 1.2.

Recently, biomass co-firing is practiced in many of the coal-based power plants to decrease the emission of SO_x, NO_x, and CO_2 per unit of electricity generated. This process has economic and environmental advantages. Along with the base fuel (i.e. coal) biomass is mixed in three different ways: (i) direct co-firing: biomass is directly fed into the boiler with coal; (ii) indirect co-firing: biomass is converted into syngas in a separate gasifier and the syngas is fed into the boiler of the base fuel; (iii) parallel co-firing: a completely separate biomass-fired boiler is used to generate steam and this steam is used to meet the demand of the coal-fired power plant (Roni et al. 2017). Many countries are adopting co-firing to meet the targets of reduction of greenhouse gas emission.

In coal or biomass or co-firing power plant, the fuel feed (say coal) is fired in the boilers of power plants to produce steam. Under tremendous pressure the steam rotates the turbine which spins the generator to produce electricity. The flowchart

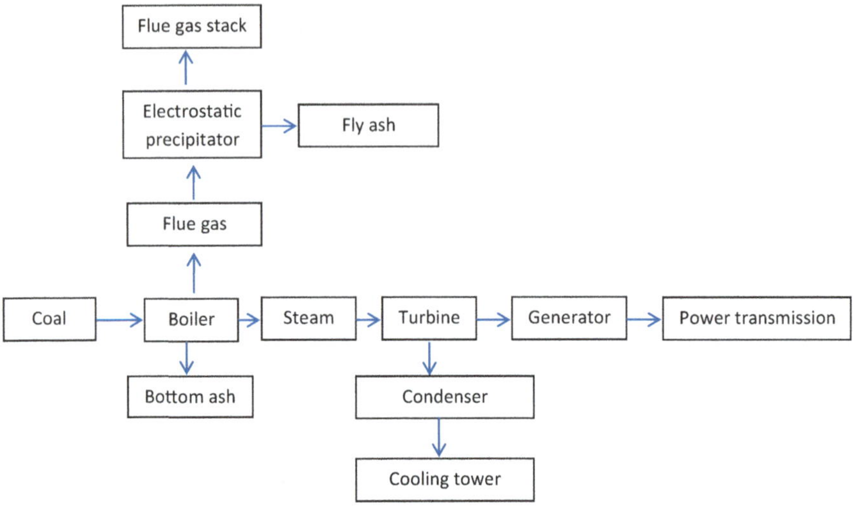

Fig. 1.2 Flowchart of a typical coal-fired thermal power plant

Table 1.3 Country-wise ash generation and utilization

Country	Ash generation	Utilization	Year	Reference
China	~600 Mt	~490 Mt	2015	(Ma et al. 2019)
India	196.44	131.87	2018–2019	(CEA 2020)
USA	102.3	59.4	2018	(ACAA 2018)

of a typical power plant is given in Fig. 1.2. The bottom ash is collected below the boiler through a water-filled hopper. Water is used to quench the clinkers and ash falling from the boiler.

In most of the power plants, the bottom and the fly ash are collected separately. The fly ash is directly shipped to the different end users. All the bottom ash and the unutilized fly ash are generally stored in large ash lagoons. Globally, 7.8 billion tons of coal ash are produced by combustion of coal in power utilities. Country-wise ash generation and utilization are presented in Table 1.3.

Worldwide, the annual generation of biomass ash is about 500 million tons (Cruz et al. 2019). The ash characteristics vary not only with the nature of biomass used but also on the combustion conditions. Thus, the operational philosophies could be managed to suit the intended utilization of the ash generated from biomass power plants. Like coal ash, both bottom and fly ashes are collected separately. Grate and fluidized bed combustions are the most common technologies. In grate combustion the efficiency is low and the ash may contain more unburnt carbon, whereas in fluidized bed combustion the ash will have low unburnt carbon but more silica due to the admixture of ash with sand used for bedding. Some biomass ash also contains significant amount of potentially toxic trace elements, so proper characterization of the ash before its utilization or disposal is important. An estimate on the biomass ash generated from different countries is presented in Fig. 1.3.

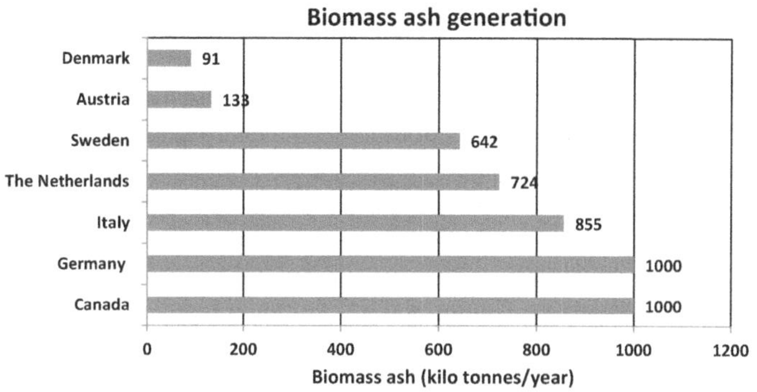

Fig. 1.3 Biomass ash produced in different countries (*Source* IEA (2018))

Table 1.4 Major areas of fly ash utilization in India (data compiled from CEA 2017, 2018, 2019–2020)

Years → Areas of utilization in % ↓↓	2016–2017	2017–2018	2018–2019
Cement	23.98	25.60	26.88
Mine filing	6.96	6.37	4.65
Bricks and tiles	8.81	9.01	9.96
Reclamation of low-lying areas	6.52	10.48	13.51
Ash dyke risings	7.02	6.90	9.94
Roads and flyovers	3.66	3.40	4.48
Agriculture	1.14	0.29	0.63
Concrete	0.45	0.66	0.82
Hydropower sector	0.01	0.004	0.00
Others	4.72	4.42	6.72

Ash handling and safe disposal is a major challenge for the coal or biomass-based power plants. Though there are few avenues for sustainable utilization of ash, significant amount of ash is unutilized and are dumped in the ash lagoons. The recent trends in ash utilization pattern of India are depicted in Table 1.4.

To increase the ash utilization, better understanding of the genesis of ash in coal and biomass is essential. Therefore, some fundamental knowledge on coal and biomass minerals is highlighted in the following session.

Evaluation of coal types involves both macroscopic and microscopic approaches. Macroscopically, two distinct and broad types of coal are identified: (a) banded coals (inhomogeneous and more common in nature) and (b) sapropelic or non-banded (homogeneous, but less common) (Taylor et al. 1998). The inhomogeneous or banded nature of the first type as mentioned above is represented in natural coal systems with existence of mainly four lithotypes (macroscopically identifiable), viz., vitrain, clarain, durain, and fusain (ICCP 1963), which are distinguished from one another in terms of color, luster, and physical properties. Microscopically, coals are composed of macerals (organic) and minerals (inorganic). Macerals, which are identified based on their optical properties in polished sections under the microscope under incident light, are distinguished into three groups: vitrinites/huminites, inertinites, and liptinites (ICCP 1963, 1971, 1975, 1998, 2001). For details on maceral groups, the types of macerals present under each group, their origin, significance, and optical properties, readers may refer to pioneering works of Stach et al. (1982), Taylor et al. (1998), and several works of International Committee of Coal and Organic Petrology (ICCP).

The combustion performance of coal is fundamentally related to all aspects of coal petrology. For example, the rank of coal and the nature of coal maceral composition control the heating value and combustion properties of coal. Different macerals behave differently (rate of reaction and quantity of heat generated) during heating in the presence of air. The influence of coal petrographic properties on its combustion was identified as early as in 1980s (Neavel 1981; Bengtsson 1986), and the main factors that were identified are vitrinite reflectance, grindability, and petrographic composition. For example, low rank and high volatile matters yielding coals ignite more easily than the higher ranked counterparts. In terms of maceral composition, it is now fairly well established that order or temperature of combustibility varies as follows: liptinite < vitrinite < inertinite. Furthermore, microlithotypes present within the coals can also affect the combustion kinetics. For example, liptinite macerals may be associated with both vitrinites and inertinites. However, its intimate association with vitrinites will help combustibility more than its association with inertinite macerals. Consequently, the reactivity of clarite lithotype towards combustion would be higher than that of durites. Inertinites, on the other hand, known to be the least reactive coal maceral component, have seldom been a point of interest among coal petrographers regarding its role in combustion. Earlier researches have documented that combustion of high inertinite bearing coals was associated with lower combustion efficiencies and the fly ashes generated to have higher unburnt carbon content (Yavorskii et al. 1968; Nandi et al. 1977). However, several other researchers have documented the reactivity of inertinite macerals, especially semifusinite. Lower reflecting inertinites, the semifusinites, with reflectance lower than 1.3% in sub-bituminous and lower rank bituminous coals from India, were observed to show encouraging reactivity and burn-out during combustion (Choudhury et al. 2008). While the origin of inertinites is itself a point of debate (see Hudspith and Belcher 2020), it is likely that different inertinite macerals would respond differently during coal combustion, and may contribute to the unburnt carbon in fly ash and bottom ash.

In addition to the organic matter in coals the inorganic materials also exist which contribute to the ash component produced during coal combustion. Common minerals in coal are presented in Table 1.5.

One of the most commonly occurring mineral in coal, quartz, fuses at exceptionally high temperatures (around 1600 °C), and thus essentially remains inert during combustion and perseveres through combustion process. However, through slow-rate solid-state reactions phases such as tridymite, cristobalite may form from quartz

Table 1.5 Most commonly occurring mineral matter in coal (Finkelman et al. 2019; Ward 2016)

Class	Common minerals	Uncommon minerals	Rare minerals
Tectosilicates	Quartz	Microcline, Orthoclase, Sanidine, Plagioclase, Albite, Anorthite, Opal	Analcime, Heulandite, Clinoptilolite
Phyllosilicates	Kaolinite, Illite	Dickite, Smectite, Montmorillonite, Halloysite, Tobelite, Pyrophyllite, Chlorite, Chamosite, Clinochlore, Muscovite	Nacrite
Inosilicate	–	Pyroxene, Augite, Amphibole	Diopside, Hornblende
Nesosilicate	Zircon	Garnet	Titanite, Grossular, Mullite
Phosphates	Apatite	Monazite, Crandallite, Goyazite, Vivianite, Rhabdophane	Vivianite
Sulfates	Barite, Jarosite	Alunite, Natroalunite, Rozenite, Hexahydrite, Melanterite, Epsomite, Pickeringite	Kieserite
Carbonates	Calcite, Siderite, Ankerite, Dolomite	Aragonite, Magnesite, Witherite, Dawsonite	Malachite, Alstonite
Oxides	Rutile, Anatase	Brookite, Hematite, Ilmenite, Magnetite	Uraninite, Spinel, Chromite, Corundum, Uraninite
Sulfides	Pyrite, Chalcopyrite, Sphalerite, Galena	Millerite, Linnaeite, Marcasite	Bornite, Argentite, Pentlandite, Pyrrhotite, Greenockite, Alabandite, Arsenopyrite
Halides	–	Halite, Sylvite	–
Fluorides	–	Fluorite	–
Hydroxide		Diaspore, Goethite, Gibbsite, Boehmite	–

(Reifenstein et al. 1999). Clay minerals, commonly occurring in coal deposits, show varied degree of phase transformations during combustion. Kaolinite transforms to metakaolinite at approximately 500 °C by losing the OH units in its structure. The metakaolinite is subsequently transformed to different phases (viz., gamma-alumina, mullite, and cristobalite) in between temperatures of 950 and 1600 °C. In the same temperature range, illitic and smectitic clays transform to spinel and mullite, and often fuse at lower temperatures to form glassy components. Beyond quartz and clay group of minerals, carbonates also occur frequently in coal deposits and show different transformation behavior when exposed to higher temperatures during coal combustion. At around 900 °C calcite gets decomposed and forms lime (CaO), while dolomite disintegrates to form CaO and periclase (MgO) in dual-step process. The advantage of calcium to incorporate sulfur during combustion to form anhydrite has been used favorably in FBC systems to minimize sulfur release to the atmosphere (Filippidis et al. 1996). Pyrite and siderite when present in coal, ideally break down to form iron oxides (viz. hematite, magnetite), and the iron often reacts to form a host of other minerals (Huffman et al. 1981). Less surprisingly, when pyrite and marcasite concentration in coal is high, their combustion will produce coal-ash with greater Fe-content. Thus, the composition of the mineral matter or inorganic components in coal is extremely critical, especially when considering the chemistry of the produced fly ash and bottom ash, and their subsequent usage at different end-industries.

Associated with the mineral matter in coal are also the 'important' trace elements (Kolker 2012; Dai et al. 2012a, b; Diehl et al. 2012). Few trace elements are also known to be intimately associated with the organic fractions of coal (Finkelman et al. 2018). Trace elements are known to be associated with certain 'organic coaly-matter' groups, viz., carboxylic-acid, phenolic-hydroxyl, mercapt and imino (Swaine 1990). Zubovic et al. (1960, 1961) observed and documented that element with smaller and highly charged ions to be allied with organic coaly-matter groups. Obviously, such associations would be highly rank dependent, as it is fairly well understood that with increasing coalification rank, functional groups are eliminated from the organic coaly-matter. However, it is fairly well accepted that majority of trace elements in coal occur within the inorganic mineral matter, either as minor or major components in minerals. Positive correlations between elemental concentrations with mineral matter concentration or ash yield enhanced trace element contents of high-density fractions washed from coal (Spears and Zheng 1999; Wagner and Tlotleng 2012), all indicates the inorganic affinity for trace elements in coal. Dai et al. (2014) in their work on coal-hosted Ge deposits from different parts of China identified Ge-oxide through XRD analysis of the coal-ash, while other Ge-bearing phases were identified through simultaneous scanning electron microscopy and energy dispersive X-ray spectroscopy of the same ashes. Other studies have also documented the presence of other strategic elements within the inorganic-fraction or minerals of coal.

Silicate minerals are undoubtedly the most common and complex group of minerals found in coal (Table 1.1). Less amusingly, the silicates in coal host several elements (viz., silicon, aluminum, potassium, calcium, sodium, magnesium, and iron; Finkelman et al. 2019). Clay minerals in coal are ideally richer in concentrations of trace and major elements, relative to other minerals. Several earlier studies have documented positive relationship between trace element concentration in coal and their clay mineral concentrations. Ideally, clay minerals, owing to their higher surface/volume ratio and negative charges, adsorb the positively charged trace elements onto their surfaces. Published reports have documented clay minerals in coal to host a variety of trace elements, viz., titanium, chromium, and so on (Ward et al. 1999; Huggins et al. 2000). Interestingly, Finkelman et al. (2019) mentioned that largely the primary elements associated with silicates are not 'critical' elements which are associated with fouling in power plant boilers, or negative for the environment and health. In contrast to the silicates, elements present within sulfides and selenides in coals are often observed to be toxic for health and environment.

Among the most common minerals in coal, pyrite of sulfide group is known to have a host of deleterious impacts in and around coal mining areas. Acid mine drainage, acid rain and associated smog, formation of slags in boiler to development of pneumoconiosis in mine workers are some of the widely known effects caused due to the presence of silica, pyrite, and coal dust (Campbell et al. 2001; Dai et al. 2002; Regina et al. 2004; Huang et al. 2006). Silica-rich constituents are also known to be a causative factor for silicosis in human beings. Host of elements, viz., sulfur, arsenic, thallium, mercury, lead, selenium, and so on, are known to be hosted by sulfides in coal (Finkelman et al. 2019). The phosphate minerals in coal are often found to be attractive in terms of the trace elements that are associated with them. Ranging from phosphorous, barium, uranium, calcium, strontium, aluminum, to rare earth elements and yttrium, phosphate minerals are known to host in variable proportions. Similarly, carbonate minerals when present in coal, are known to be 'good' hosts for a range of trace elements viz. calcium, magnesium, iron, manganese, strontium and REE (Finkelman et al. 2018, 2019).

The trace elements in coal often are found to be concentrated in the ashes derived from coal combustion (Querol et al. 1995; Meij and teWinkel 2009; Vassilev et al. 2009). In as early as 1938, Sinnatt and Baragwanath (1938) detected trace element concentration in ashes from Victorian coals, Australia. It is now also widely established that minerals in coal are the chief contributor toward ash (fly and bottom) generation, and boiler slag formation in thermal power plants. Proper extraction of important components and strategies to use these as value-added materials add substantial costs to the use of coal. Presence of several critical elements, viz., REE and Y, Li, Ga, Se, and so on has attracted the attention of researchers toward coal and coal ash for its specific usage in semi-conductor industries. Further, several studies have also revealed economic concentrations of REE in coal ashes, thus making it an economically interesting option (Seredin and Dai 2012; Wagner and Matiane 2018). Higher alumina concentrations in coal ash, coming from clay minerals predominantly, have also received attention due to the possibilities of Al extraction from them (Seredin 2012). All these indicate the importance of coal ash and strategically puts forward the importance of ash studies and ash utilization initiatives.

Beyond being the immediate source of several critical elements, the fly ash and bottom ash produced from coal combustion, that is, coal combustion by-products are now being seen as reliable and valued commodities in different sectors. Due to high usage of coal in India, China, and Poland, and consequently generation of greater amounts of fly ash and bottom ash, these countries now have a surplus of this by-product, which can act as a reliable 'stock'/product supply to other countries for usage in different sectors. In India, the strategic importance of these valued commodities is very well understood, and consequently, strong plans have been proposed by the Government of India for its strategic utilization.

The origin of coal, ash, or mineral matter composition of the coal decides the nature and properties of the fly ash generated in power plants. For instance, Indian coals are marked by higher ash yields, and consequently large quantities of fly and bottom ash are generated in the thermal power plants of the country. The Indian

Government was quick to understand the major issues on fly ash storage in suitable 'land' areas and the environmental concerns in relation to fly and bottom ash generation in huge quantities (CEA 2017). In order to address these, the Ministry of Environment, Forests and Climate Change (MoEF & CC), Government of India had released several announcements on fly ash utilization from 1999 to 2009 with the main objective of creating a system which allows 100% utilization of ash generated in a phased-manner. Further, several deliberations, discussions, and strong monitoring were and are being currently adopted by the Ministry of Environment, Forests and Climate Change (MoEF & CC) for proper fly ash utilization and reducing any form of associated environmental risks. In their 2016 report (Report No. 39 of 2016), the Ministry of Environment, Forest and Climate Change, Government of India, through their exhaustive endeavor to thoroughly monitor all thermal power check-points for fly ash utilization, identified several states in India where either of storage, utilization (within a radius of 300 km from the TPPs), or fugitive ash emission from TPPs were not being followed carefully. Consequently, a huge improvement in fly ash utilization has been observed in India in the last few years. While in 2016–2017, fly ash generation, utilization, and percentage utilization stood at 169.25 million tons, 107.10 million tons, and 63.28%, respectively, in 2017–2018 the same figures stood at 196.44 million tons, 131.87 million tons, and 67.13%, respectively (CEA 2017, 2018). The figures showed even more improvement in 2018–2019 with generation, utilization, and percentage utilization of fly ash standing at 217.04 million tons, 168.40 million tons, and 77.59%, respectively (CEA 2020). The improvement in percentage utilization stands out even more when the old figures of 1999–2000 are considered. In 1999–2000, generation, utilization, and percentage utilization of fly ash stood at 74.03 million tons, 8.91 million tons, and 12.03%, respectively. The major areas of fly ash utilization in India (Table 1.2) have been observed to be cement sector (as pozzolanic material, and thereby reducing percentage consumption of limestone and coal) followed by those used in producing bricks and tiles, ash dyke raising, mine filing and reclamation of low-lying areas (as a substitute of top soil and sand), in roads and embankments, agriculture (due to its content of micro-nutrients), concrete production, and hydropower sectors (CEA 2017, 2018, 2020). However, for achieving the targets of 100% utilization, more concerted efforts are needed in the coming years.

In the USA, beyond the usage of the coal ash in different sectors (viz., construction, cement, etc.), several norms are placed for secured utilization/clearance of coal-combustion residues. A range of factors, viz., locational restrictions, design parameters (for preventing leaching from stored ash to groundwater), structural integrity, operating criterion, groundwater chemistry and monitoring, closure/post-closure requirements, are considered. The laws and regulations pertaining to coal ash use, certification, and management in the USA are dissimilar at different states. Generally, certification of coal ash employs evaluation of the character of both the ash and the mine site. Only after these thorough checking, the ash usage plan may be implemented. Beyond the above-mentioned two criteria, the quality of groundwater and surface water is thoroughly inspected both at downstream and upstream positions of the mine, and proper plans are employed for meeting 'safe' drinking water requirements of the specific state. Similar to the USA, in Australia, different states use different standards for the usage of coal ash. In Queensland, with the target of improved waste management, in 2019 the state regulations were revised and coal combustion products were categorized as resource materials and not wastes. However, the coal ash could be used as a resource by the producer, only after stringent monitoring and analysis of the ash, and only if it falls in accordance with this end-of-waste code. Additionally, the ash must not be marked by the presence of any contaminants, at concentrations which may cause environmental harm. In China, huge amount of ash is produced annually as coal happens to be the main-stay of their energy economy. Consequently, a lot of stress is laid upon the ash generated from coal, and has been categorized as '*Class 2 General Industrial Solid Waste*'. The Chinese Government has been pushing for a comprehensive utilization policy for coal ash, and has introduced several sustainable guidelines for environmental safety. Currently, the ash produced in China is being used in mine backfilling, soil improvement and fertilizer promotion, cement industries, road, embankment, and other concrete structures.

Though biomass is the precursor of coal, but due to the various geological factors of coal formation, the ash composition of coal and biomass is distinct. Coal ash is dominated by alumina silicates, whereas biomass ash by alkalies. This distinct chemical composition has significant influence on the ash handling and utilization. The existing knowledge base on the coal ash has to be adopted for biomass ash. Many countries are in transition to switch from coal-based energy to biomass-based renewable energy sources. Biomass co-firing and standalone biomass power plants are being practiced. Comparative evaluation of coal and biomass ash in terms of their properties to potential utilization is the key novel feature of this monograph. Though these two fuel feedstocks originated from plants, the environmental conditions that prevailed during their genesis and further transformation had significantly affected the nature of the mineral matter in the feedstock, and accordingly the ash generated after combustion has distinct characteristics. Coal ash is primarily composed of oxides of Si, Al, Fe, and Ca, whereas Si, Ca, K, Na are the dominant constituents of biomass ash. The mineral content in coal can vary largely depending upon their mode of origin and deposition, whereas biomass has low ash content due to the phyto origin of the inorganic constituents. Though coal and biomass are closely related, one-stop literature on their ash is not available. Thus, the primary objective of the monograph is to present a comparative fundamental knowledge on origin of ash forming minerals in coal and biomass, their behavior during combustion, and their influence on the ash properties, ash utilization prospects, and the environmental issues.

References

ACAA (2018) ACAA 2018 CCP, Survey Results and Production & Use Charts. American Coal Ash Association

Bengtsson M (1986) Combustion behavior for a range of coals of various origins and petrographic composition. PhD dissertation, Stockholm, Sweden, The Royal Institute of Technology, variously paginated

Campbell RN, Lindsay P, Clemens AH (2001) Acid generating potential of waste rock and coal ash in New Zealand coal mines. Int J Coal Geol 45:163–179

CEA (2017) CEA, Annual Report 2016-17. Central Electricity Authority Ministry of Power, Government of India

CEA (2018) CEA, Annual Report 2017-18. Central Electricity Authority Ministry of Power, Government of India

CEA (2020) CEA, Annual Report 2018-19. Central Electricity Authority Ministry of Power, Government of India

Choudhury N, Biswas S, Sarkar P, Kumar M, Ghosal S, Mitra T, Mukherjee A, Choudhury A (2008) Influence of rank and macerals on the burnout behaviour of pulverized Indian coal. Int J Coal Geol 74:145–153

Clarke L, Jiang K, Akimoto K, Babiker M, Blanford G, Fisher-Vanden K, Hourcade J-C, Krey V, et al (2014) Assessing transformation pathways. In: Edenhofer O et al (eds) Climate change 2014: Mitigation of climate change. contribution of working group III to the Fifth assessment report of the intergovernmental panel on climate change. Cambridge Univ Press, Cambridge, UK, pp 413–510

Commission of the European Communities (2005) Annex to the Communication from the Commission to the Council and the European Parliament on Community Strategy Concerning Mercury, extended impact assessment, pp 174. http://europa.eu.int/comm-/environment/chemicals/mercury/index.htm and http://ec.europa.eu/environment/-chemicals-/mercury/pdf/extended_impact_assessment.pdf

Cruz NC, Silva FC, Tarelho LA, Rodrigues SM (2019). Critical review of key variables affecting potential recycling applications of ash produced at large-scale biomass combustion plants. Resour, Conserv Recycl 150:104427

Dai S, Ren D, Tang Y, Shao L, Li S (2002) Distribution, isotopic variation and origin of sulfur in coals in the Wuda Coalfield, Inner Mongolia, China. Int J Coal Geol 51:237–250

Dai S, Jiang Y, Ward CR, Gu L, Seredin VV, Liu H, Zhou D, Wang X, Sun Y, Zou J, Ren D (2012a) Mineralogical and geochemical compositions of the coal in the Guanbanwusu Mine, Inner Mongolia, China: further evidence for the existence of an Al (Ga and REE) ore deposit in the Jungar Coalfield. Int J Coal Geol 98:10–40

Dai S, Ren D, Chou C-L, Finkelman RB, Seredin VV, Zhou Y (2012b) Geochemistry of trace elements in Chinese coals: a review of abundances, genetic types, impacts on human health, and industrial utilization. Int J Coal Geol 94:3–21

Dai S, Li T, Seredin VV, Ward CR, Hower JC, Zhou Y, Zhang M, Song X, Song W, Zhao C (2014) Origin of minerals and elements in the Late Permian coals, tonsteins, and host rocks of the Xinde Mine, Xuanwei, eastern Yunnan, China. Int J Coal Geol 121:53–78

Diehl SF, Goldhaber MB, Koenig AE, Lowers HA, Ruppert LF (2012) Distribution of arsenic, selenium, and other trace elements in high pyrite appalachian coals: evidence for multiple episodes of pyrite formation. Int J Coal Geol 94:238–249

Edenhofer O, Pichs-Madruga R, Sokona Y, Seyboth K, Matschoss P, Kadner S, von Stechow C (2011) Renewable energy sources and climate change mitigation. Cambridge University Press, Cambridge

Filippidis A, Georgakopoulos A, Kassoli-Fournaraki A (1996) Mineralogical components of some thermally decomposed lignite and lignite ash from the Ptolemais basin, Greece. Int J Coal Geol 30:303–314

Finkelman RB, Palmer CA, Wang P (2018) Quantification of modes of occurrence of 42 elements in coal. Int J Coal Geol 185:138–160

Finkelman RB, Dai S, French D (2019) The importance of minerals in coal as the hosts of chemical elements: a review. Int J Coal Geol 212:103251

Huang X, Gordon T, Rom WN, Finkelman RB (2006) Interaction of iron and calcium minerals in coals and their roles in coal dust-induced health and environmental problems. Rev Mineral Geochem 64:153–178

Hudspith VA, Belcher CA (2020) Some semifusinite in coal may form during diagenesis, not wildfires. Int J Coal Geol 218:103360

Huffman GP, Huggins FE, Dunmyre GR (1981) Investigation of the high temperature behaviour of coal ash in reducing and oxidizing atmospheres. Fuel 60:585–597

Huggins FE, Shah N, Huffman GP, Kolker A, Crowley SS, Palmer CA, Finkelman RB (2000) Mode of occurrence of chromium in four U.S. coals: in review, 'toxic Substances from Coal Combustion'. Fuel Process Technol 63:79–92

IEA (2018) International Energy Agency (IIEA) Bioenergy Task 32. Options for increased use of ash from biomass combustion and co-firing. Published by IEA Bioenergy

International Committee for Coal Petrology (ICCP) (1963) International handbook of coal petrography, 2nd edn. CNRS (Paris)

International Committee for Coal Petrology (ICCP) (1971) International handbook of coal petrography, 1st supplement to 2nd edn. CNRS (Paris)

International Committee for Coal Petrology (ICCP) (1975) International handbook of coal petrography, 2nd supplement to 2nd edn. CNRS (Paris)

International Committee for Coal and Organic Petrology (ICCP) (1998) The new vitrinite classification (ICCP System 1994). Fuel 77:349–358

International Committee for Coal and Organic Petrology (ICCP) (2001) The new inertinite classification (ICCP System 1994). Fuel 80:459–471

Irena (2012). Renewable energy technologies: cost analysis series. Wind power. Volume 1: Power sector Issue 5/5. Biomass for power generation. International Renewable Energy Agency. IRENA

Irena (2018) Renewable capacity statistics 2018, International Renewable Energy Agency (Irena), Abu Dhabi

Kolker A (2012) Minor element distribution in iron disulfides in coal: a geochemical review. Int J Coal Geol 94:32–43

Lu X, Cao L, Wang H, Peng W, Xing J, Wang S et al (2019) Gasification of coal and biomass as a net carbon-negative power source for environment-friendly electricity generation in China. Proc Natl Acad Sci USA 116:8206–8213

Ma S-H, Xu M-D, Qiqige X, Wang X-H, Zhou X (2019) Challenges and developments in the utilization of fly ash in China. Int J Environ Sci Develop 8(11)

Masson-Delmotte V (2018) An IPCC special report on the impacts of global warming of 1.5 °C above pre-industrial levels and related global greenhouse gas emission pathways, in the context of strengthening the global response to the threat of climate change, sustainable development, and efforts to eradicate poverty. Intergovernmental Panel on Climate Change, Geneva

Meij R, te Winkel BH (2009) Trace elements in world steam coal and their behaviour in Dutch coal-fired power stations: A review. Int J Coal Geol 77:289–293

Nandi BN, Brown TD, Lee GK (1977) Inert coal macerals in combustion. Fuel 56(2):125–130

Neavel RC (1981) Origin, petrography and classification of coal. In: Elliott MA (ed) Chemistry of coal utilization, 2nd Supplementary Volume. John Wiley and Sons, New York, pp 91–158

Querol X, Fernandez-Turiel J, Lopez-Soler A (1995) Trace elements in coal and their behaviour during combustion in a large power station. Fuel 74:331–343

Regina JR, DuPont JN, Marder AR (2004) Corrosion behavior of Fe-Al-Cr alloys in sulfur-and oxygen-rich environments in the presence of pyrite. Corrosion 60:501–509

Reifenstein AP, Kahraman H, Coin CDA, Calos NJ, Miller G, Uwins P (1999) Behaviour of selected minerals in an improved ash fusion test: quartz, potassium feldspar, sodium feldspar, kaolinite, illite, calcite, dolomite, siderite, pyrite and apatite. Fuel 78:1449–1461

Roni MS, Chowdhury S, Mamun S, Marufuzzaman M, Lein W, Johnson S (2017) Biomass co-firing technology with policies, challenges, and opportunities: a global review. Renew Sustain Energy Rev 78:1089–1101

Seredin VV (2012) From coal science to metal production and environmental protection: a new story of success. Int J Coal Geol 90–91:1–3

Seredin VV, Dai S (2012) Coal deposits as potential alternative sources for lanthanides and yttrium. Int J Coal Geol 94:67–93

Sinnatt FS, Baragwanath GE (1938) The hydrogenation of Victorian brown coals. Rep Fuel Res Station HI:349–503

Stach E, Mackowsky M-Th, Teichmüller M, Taylor GH, Chandra D, Teichmüller R (eds) (1982) Stach's textbook of coal petrology. Gebrüder Borntraeger, Berlin

Spears DA, Zheng Y (1999) Geochemistry and origin of elements in some UK coals. Int J Coal Geol 38:161–179

Swaine DJ (1990) Trace elements in coal. Butterworths, London, p 278

Taylor GH, Teichmüller M, Davis A, Diessel CFK, Littke R, Robert P (1998) Organic petrology. Gebrüder Borntraeger, Berlin

U.S. Environmental Protection Agency (US-EPA) (2005) Clean air interstate rule. www.epa.gov/interstateairquality/

Vassiliev SV, Vassileva CC, Baxter D, Andersen LK (2009) A new approach for the combined chemical and mineral classification of the inorganic matter in coal. 2: potential applications of the classification systems. Fuel 88:246–254

Wagner NJ, Tlotleng MT (2012) Distribution of selected trace elements in density fractionated Waterberg coals from South Africa. Int J Coal Geol 94:225–237

Wagner NJ, Matiane A (2018) Rare earth elements in select Main Karoo Basin (South Africa) coal and coal ash samples. Int J Coal Geol 196:82–92

Ward CR (2016) Analysis, origin and significance of mineral matter in coal: an updated review. Int J Coal Geol 165:1–27

Ward CR, Spears DA, Booth CA, Staton I, Gurba LW (1999) Mineral matter and trace elements in coals of the Gunnedah Basin, New South Wales, Australia. Int J Coal Geol 40:281–308

Ward CR, Suárez-Ruiz I (2008) Introduction to applied coal petrology. Applied coal petrology, pp 1–18. https://doi.org/10.1016/b978-0-08-045051-3.00001-4

World Coal Institute (2005) The coal resource: a comprehensive overview of coal. World Coal Institute, Richmond, UK, pp 44. www.worldcoal.org/assets_cm/files/PDF/thecoal-resource.pdf

www.bp.com (2019) Statistical Review of World Energy

Yavorskii IA, Alaev GP, Pugach LI, Talankin LP (1968) Influence of the petrographic composition of coals on the efficiency of a pf fired boiler furnace. Теплоэнергетика 9:69–72

Zubovic P, Stadnichenko T, Sheffy NB (1960) The association of minor elements with organic and inorganic phases in coal. U.S. Geol Surv Prof Pap 400-B:B84–B87

Zubovic P, Stadnichenko T, Sheffy NB (1961) The association of minor elements associations in coal and other carbonaceous sediments. U.S. Geol Surv Prof Pap 424-D(411):D345–D348

Chapter 2
Genesis and Characteristics of Coal and Biomass Ash

Abstract The coal ash and biomass ash produced from their respective combustion are widely variable in terms of their physical and chemical properties. Thorough understanding of their properties is a prerequisite for their possible utilization at different sectors. In this chapter we discuss different physical and chemical parameters, viz., particle size distribution, surface area and porosity, major oxide composition, percentage of unburned carbon, and micro-morphology of several ash samples from India. Presence of unburnt carbon, that is, non-combusted or partially combusted organic matter in ash, can cause potential damage when used in concrete structures, as it induces adsorption of air-entraining agents (or AEAs) and thereby limits their role in resisting freeze-and-thaw in concretes. Consequently, knowledge on unburnt carbon, surface area, and porosity of ashes can be critical. Similarly, the chemical composition of the ashes is significantly important, and depending upon their composition, the ashes can be used in relevant sectors. The coal ashes were observed to be marked by higher $SiO_2 + Fe_2O_3 + Al_2O_3$ contents, while the biomass ashes were observed to be marked by higher Na_2O, K_2O, MgO, and P_2O_5 contents.

2.1 Generation of Coal Combustion Residues

Predominant products formed from coal combustion are depicted in Fig. 2.1. Fly ash (FA), which happens to be one of the most strategically significant coal combustion products (CCPs), is accumulated by mechanical filters or electrostatic precipitators from the flue gas produced during coal combustion. The nature of parent coal strongly controls the nature of the FA produced. Generally, combustion of low-rank coals (lignites and sub-bituminous coals) yields Class C-ash, while high-rank coals (bituminous and anthracites) generates Class F-ash. The properties of the above-mentioned classes of ash are discussed later in this chapter. As coal is introduced into the boiler in power plants, the different stages of coal combustion that take place are as follows: (a) first stage is removal of moisture; (b) stage 'a' followed by combustion of volatile matter (VM) in coals; (c) stages 'a' and 'b' followed by the

A. K. Singh et al., *Ash from Coal and Biomass Combustion*,
https://doi.org/10.1007/978-3-030-56981-5_2

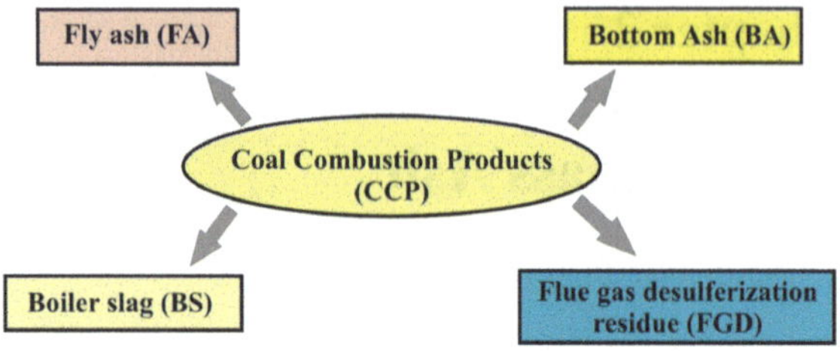

Fig. 2.1 Predominant products formed from coal combustion

stage where combustion of fixed carbon (FC) in coal takes place. After the combustion of the coal, non-combustible inorganic phases are left behind. In this chapter we show the different laboratory characterization techniques of fly ash, bottom ash, and biomass.

Bottom ash (BA), which is collected from the bottom part of boilers in thermal power plants, represents the uncombusted materials produced from coal combustion; boiler-slags are commonly formed when the operation temperature in the boilers exceed the fusion temperatures of ash. The molten-slag is drained from the basal part of the combustion chamber (Sajwan et al. 2006). Morphologically, these bottom ashes are grainy to granular, mimicking the appearance of concrete sands (Keefer 1993), and show abrasive properties finding their usage in structural embarkments and road-base matter. The usage of bottom ash is dependent on its particle or grain size distribution and color (Kula et al. 2002).

The FGD essentially became effective after the drive from environmental agencies to mitigate or restrict the release of SO_x from TPPs, before which most companies utilized high sulfur-bearing bituminous coals. The main motive of creation of flue gas scrubbing facilities at TPPs leading to FGD was to reduce pollution from coal combustion in coal-fired plants. FGD residue represents the alkaline substance formed when SO_x is removed from thermal power plant flue gases (Pushon et al. 2002). These typically are composed of calcium sulfate ($CaSO_4$) and sulfite ($CaSO_3$), some sorbents, and fly ash components, and occasionally Mg, Al, and sodium sulfites and/or sulfates.

2.2 Generation of Biomass Ash

It is now widely recognized that biomass, due to its renewability, when combusted, affects the environment to a much lesser extent due to CO_2 neutral-conversion (Vassilev et al. 2013). Consequently, recent years have seen global attention and tremendous curiosity toward the usage of biomass as biofuels, as an alternative to fossil fuels (Vassilev et al. 2010, 2012). It has been estimated that by the year 2050 between 33 and 50% of global energy consumption could be met by biomass (Demirbas 2001; McKendry 2002; Williams et al. 2012). Broadly, combusting new biomass doesn't add-up CO_2 to the atmosphere, as re-plantations allow CO_2 to be absorbed and given back for new growth-cycle. McKendry (2002) rightly pointed out that one of the most critical factors which are neglected in the biomass CO_2 neutral-conversion concept is that a huge time-lag exists between biomass or fossil fuel combustion induced CO_2 release, and eventual biomass absorbance of CO_2. Consequently, when considering the biomass combustion, strategies should be adopted to recognize and minimize this time-lag.

Biomass combustion, which involves a set of highly complex physical and chemical processes, can be divided into initial drying stage, followed by pyrolysis stage, and a final combustion stage; and depending on the adopted technologies, the nature of these stages varies. The ash produced from thermal combustion of biomass is essentially a solid residue left behind, and as a concept is marked by wide range of macro- and micro-nutrients that are left behind after combustion. With the shifting focus toward biomass-based energy, a huge challenge in the form of large quantity of ash from biomass is being faced. The ash produced from wood or other type of biomass is often seen to vary from as low as 2 wt% (willow wood) to as high as 20 wt% (rice husks) (Jenkins et al. 1998). The biomass combustion residues vary considerably in terms of their physical and chemical characters, and it is these characteristics that determine its usage and suitability for usage at different sectors (Karltun et al. 2008). Treatment of these ashes prior to any application is extremely critical, as even adding untreated biomass ash to soil may cause significant damages to plants. The application form of biomass ash is of great concern, as untreated ash is difficult to apply anywhere, and even if added to soil directly, it may cause damage to the plant surfaces. Thus pre-treating ash to lower its reactivity and chemically reactive nature is extremely vital.

Biomass contains a variety of inorganic metals and non-metals, viz., silicon, calcium, magnesium, potassium, sodium, phosphorous, sulfur, chlorine, aluminum, iron, manganese, copper, zinc, cobalt, molybdenum, arsenic, nickel, chromium, lead, vanadium, mercury, which are present in varying concentrations, and generally act as nutrients in the plants. These elements are often concentrated in the ashes produced by biomass combustion, and despite their value, they are often disposed in landfills.

Combustion of biomass at plants produces: (a) biomass bottom ash—ash fraction created in the grate and the main combustion chamber; (b) biomass cyclone fly ash—fine particles in secondary combustion zone; and (c) biomass filter fly ash—finest biomass ash fraction formed in the electrostatic filters.

2.3 Characterization of Coal and Biomass Ash

2.3.1 Particle Size Distribution

Particle size of coal ash and its determination are extremely important for both estimating the combustion competence and for gauging its utilization possibility. As mentioned earlier, the ash generated from coal, especially the fly ash portion, is used by the industry for a range of purposes (Potgieter et al. 2005). For utilization in different sectors, pozzolanic property and the particle size of the ash being used are extremely critical, and often it has been observed that pozzolanic properties increase with decreasing particle size of ash, as lowering of particle sizes allows greater availability of silanol groups (Kumar et al. 2007).

The particle sizes of fly and bottom ashes depend on the combustion efficiency. In general, sizes range between 0.02 and 0.2 μm when formed from homogeneous-condensation, while those formed during inorganic matter fragmentation ranges between 0.2 and 10 μm (Tomeczek and Palugnoik 2002). Fly ashes with particle diameter <5–10 μm have found extensive usage in high-performance concretes, due to its capability of improving the performance of such concretes, through: (a) decreasing water 'need' of the concrete (Sun et al. 2003), (b) fortifying the 'Transition Zone' (Kuroda et al. 1993), (c) reducing 'bleeding' during concrete-mixing (Brown 1982), (d) enhancing compressive strength of the prepared concrete. Further, fine fly ashes also boost mechanical vigor of rubbers and polymers when added as fillers. Beyond the size of ash generated, their shape also places an important role in their utilization strategies. Kamruzzaman et al. (2003) documented the importance of 'ball bearing effect' of spherical-fly ash particles, which helps in improving the workability during preparation of concrete.

The NTPC thermal power plants (TPPs) in India are generally super-performers showing high combustion efficiency and generally producing finer-sized ashes (Sarkar et al. 2005). However, TPPs with smaller capacities in general are marked by poorer efficiencies and generation of coarser-sized ashes, having higher LOIs. As utilization potential is directly dependent on the particle sizes of the ashes, it is imperative that the sizes are determined cautiously. Most commonly, PSD of ashes are determined using laser particle size analyzer, which employs diffraction of laser light by the ash constituents suspended in a variety of mediums. The nature of diffraction intensity is reliant on the ash particle sizes.

Sarkar et al. (2005), in their work on particle size distribution of fly ash produced from different power plants in India, observed in the fly ash samples from Farakka Super Thermal Power Plant in Murshidabad, West Bengal, India, single bell-shaped particle size distribution curve, with 95 vol.% of the particles falling within the 1–60 μm size range. In contrast, for fly ash samples from Bokaro Thermal Power Plant and Dishergarh Power Supply Corporation, double bell-shaped curves were observed. Although not the sole criteria, earlier studies suggest that if particle diameter of ash samples is greater than 50 μm, then they are not very useful for construction purpose.

Particle size distribution data of few biomass ash samples are presented in Table 2.1. A wide variation in particle size distribution can be seen for the biomass ash samples. A glance through the data reveals that for the coconut waste biomass ash (Fig. 2.2) and mustard stalk biomass ash, the D50 is at 27.35 and 24.65 μm, and consequently they may be used as partial cement replacement for concrete. On the other hand, for the ply-wood, matchbox, and wood-waste biomass ash (Fig. 2.3), the D50 is placed at 476.57 μm, indicating its coarser nature and therefore its lack of suitability for being applied for cement replacement.

Table 2.1 Particle size distribution of few biomass ash samples from India

Si. no	Sample	d (0.1)	d (0.5)	d (0.9)
1	Paddy husk ash	9.01	40.67	103.75
2	Coconut waste ash	5.44	27.35	98.02
3	Wood waste ash	43.36	476.57	1032.22
4	Mustard stalk ash	5.14	24.65	102.89
5	Coriander ash	11.51	70.27	231.86
6	Spices and condiments ash	33.87	110.78	396.37

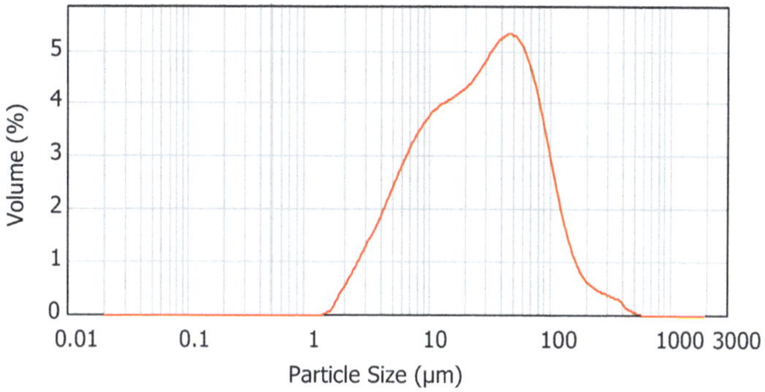

Fig. 2.2 Particle size distribution for coconut waste biomass ash with D50 at 27.35 μm

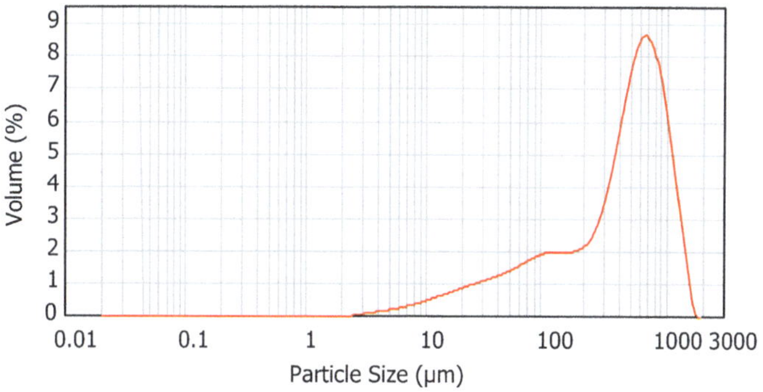

Fig. 2.3 Particle size distribution for ply-wood, matchbox, and wood-waste biomass ash with D50 at 476.57 μm

2.3.2 Surface Area and Porosity

The specific surface area (SSA) and pore size distribution (PSD) of coal ash has often been referred by researchers as a critical factor that possibly affects its utilization plan. Presence of unburnt carbon, that is, non-combusted or partially combusted organic matter in ash, causes adsorption of air-entraining agents (or AEAs) which play a critical role in resisting freeze-and-thaw in concretes (Külaots et al. 2004). The adsorption properties of organic matter have been critically studied by researchers involved in coal bed methane, shale gas/oil, CO_2 sequestration, and also in coal carbonization, combustion, and gasification (Thomas and Damberger 1976; Mahajan 1991; Hazra et al. 2018a, b, 2019a, b; Vishal et al. 2013, 2015, 2019). For understanding the pore properties, viz., pore size distribution (PSD), pore sizes, fractals, and so on, low-pressure gas adsorption technique using N_2 has been widely used by researchers (Sing 1985; Ross and Bustin 2009; Passaras et al. 2017). In this work we present the above-mentioned pore properties of few fly ash, bottom ash, and biomass ash samples.

Figures 2.4 and 2.5 represent dV/d log (r) versus pore size plots of coal fly ash samples collected from two different thermal power plants in India, viz., NTPC Vindhyachal Super Thermal Power Station (VSTPS) and NTPC Singrauli Super Thermal Power Plant (SSTPS). Less surprisingly, in both the samples, the BET SSA was observed to be low (1.27 and 0.68 m^2/g). Although the volume of pores present within the two samples is low and nearly similar (0.001774 and 0.001001 cc/g), the disposition of the pore sizes shows an increasing trend from lower pore sizes to coarser pore sizes for the fly ash sample from NTPC-VSTPS, while showing a reverse trend for the fly ash sample from NTPC-SSTPS. Consequently, the BJH average pore radius was observed to be 45.86 Å for the former and 27.77 Å for the latter. Although the fly ash sample from NTPC-SSTPS has lower pore volume compared to that of the fly ash sample from NTPC-VSTPS, the BET SSA was observed to be higher for the former. This is essentially caused due to greater volume of pores below 40 Å radius for the ash sample from NTPC-SSTPS. For the ash sample from NTPC-VSTPS a smaller volume of pores exists at the smaller size ranges, while a larger volume of pores exists at the coarser pore sizes, thereby yielding smaller BET SSA but higher BJH pore volume.

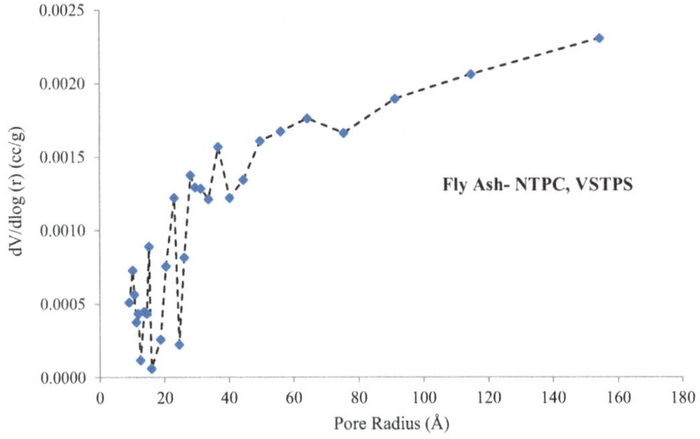

Fig. 2.4 Pore size distribution plot of fly ash sample from NTPC-VSTPS, derived from the nitrogen adsorption isotherm using BJH model

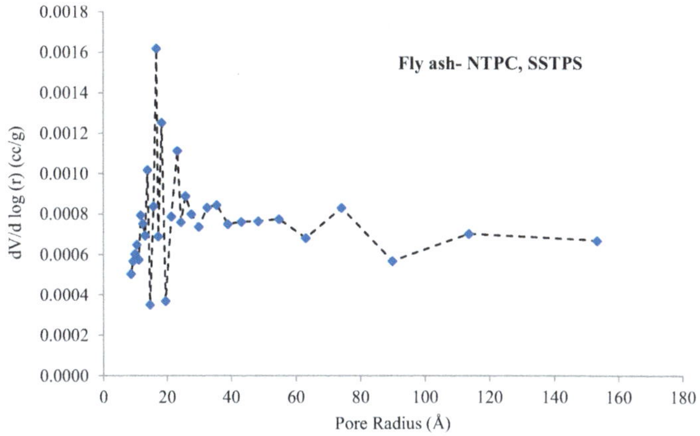

Fig. 2.5 Pore size distribution plot of fly ash sample from NTPC-SSTPS, derived from the nitrogen adsorption isotherm using BJH model

In contrast to the coal fly ashes, the bottom ashes from the same power plants showed slightly higher BJH pore volumes (0.002055 and 0.001816 cc/g, respectively) and BET SSA (1.81 and 0.91 m^2/g, respectively), which are essentially caused due to comparatively greater concentration of pores in the smaller pore size range. Figures 2.6 and 2.7 show the BJH pore size distribution plots of the two bottom ash samples measured from nitrogen gas adsorption data. The nature of disposition of the pore sizes, that is, increasing trend from smaller pores to larger pores indicates the presence of coarser pores and larger average pore sizes (29.25 and 30.91 Å) in both the bottom ashes.

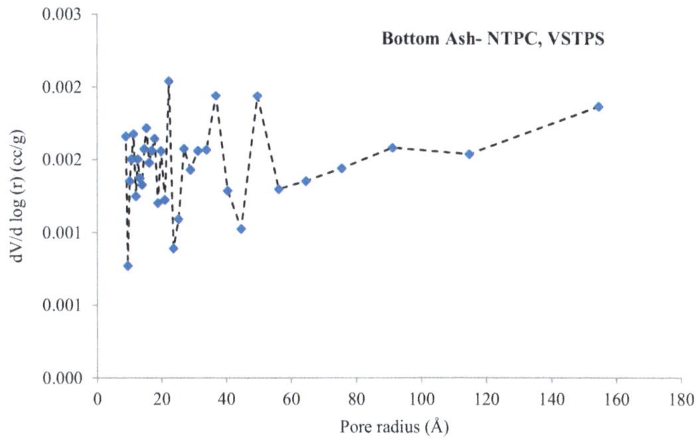

Fig. 2.6 Pore size distribution plot of fly ash sample from NTPC-VSTPS, derived from the nitrogen adsorption isotherm using BJH model

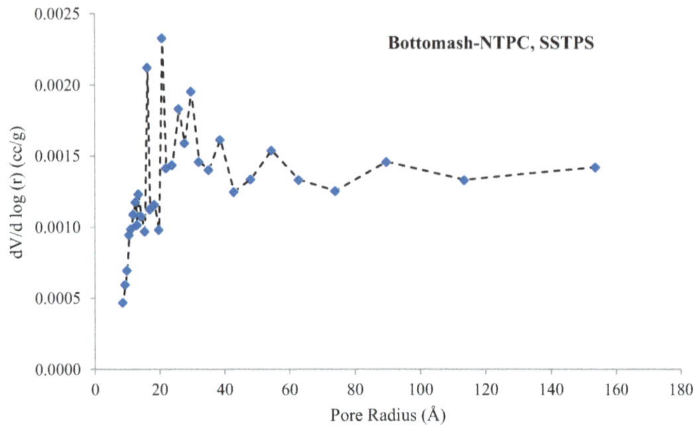

Fig. 2.7 Pore size distribution plot of bottom ash sample from NTPC-SSTPS, derived from the nitrogen adsorption isotherm using BJH model

In contrast to the coal fly ash and bottom ash samples, the lignite fly ash and bottom ash samples collected from Barmer lignite deposit under leasehold of JSW Energy (Rajasthan, India) showed higher BET SSA and much larger pore volumes. Figures 2.8 and 2.9 plot the BET SSA for the lignite fly ash and bottom ash samples. In both the cases, the R^2 was observed to be >0.999. The BET SSA for the lignite fly ash and bottom ash samples was observed to be 4.24 and 3.72 m^2/g, respectively. In contrast to the pore size distributions shown by the coal fly ash and bottom ash samples, the lignite fly ash and bottom ash samples showed distinctive plots. Figures 2.10 and 2.11 show the PSD plots of the lignite fly ash and bottom ash samples, respectively, determined using BJH model utilizing N_2 gas adsorption data. For both the lignite fly ash and bottom ash samples, an increasing trend in pore

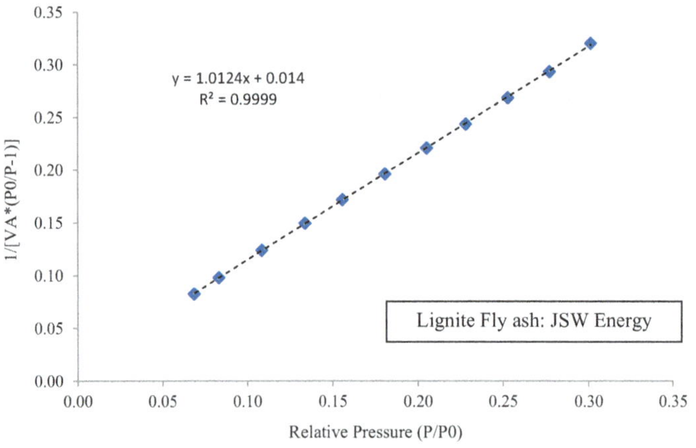

Fig. 2.8 BET-SSA plot of the lignite fly ash sample

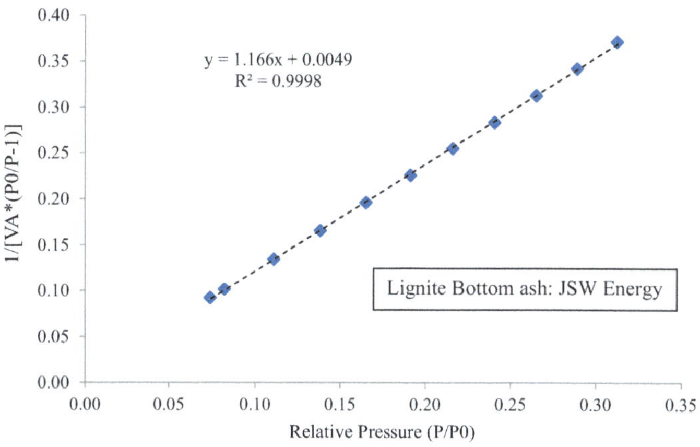

Fig. 2.9 BET-SSA plot of the lignite bottom ash sample

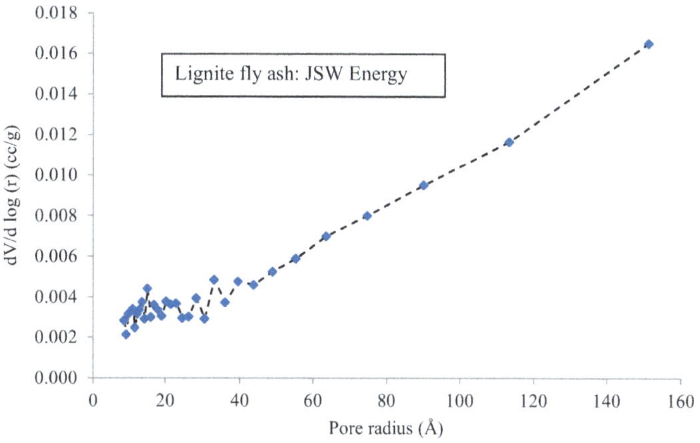

Fig. 2.10 Pore size distribution plot of lignite fly ash sample from JSW Energy, derived from the nitrogen adsorption isotherm using BJH model

size disposition from smaller to larger pores has been observed. Consequently, the average pore radius of both the lignite ash samples was observed to be much larger compared to that of the coal ash samples (43.67 and 38.88 Å). The results indicate that the lignite ash samples have substantial amounts of pores across different sizes compared to those of coal ash samples.

Pore size distribution of two biomass ash samples is shown in Figs. 2.12 and 2.13. For the spent wash sludge biomass ash (Fig. 2.12) lower BJH pore volume (0.002172 cc/g) and smaller BET SSA (1.41 m^2/g) were observed. However, for the mustard stalk biomass ash, slightly higher pore volumes (0.005463 cc/g) and BET SSA (3.42 m^2/g) were noted. Similar to the other samples, here also it can be observed that the pore size disposition shows an increasing trend from smaller pore

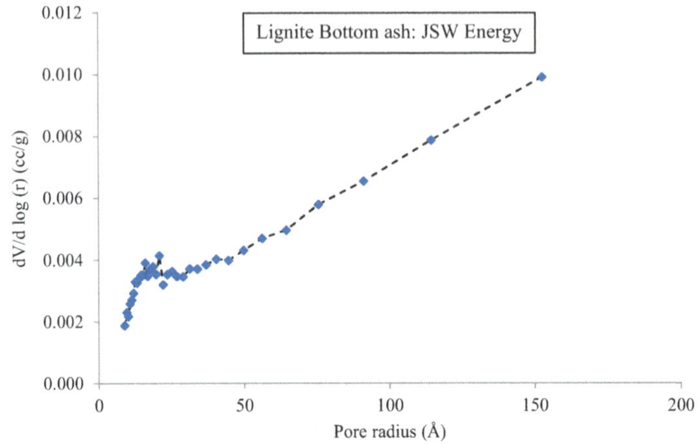

Fig. 2.11 Pore size distribution plot of lignite bottom ash sample from JSW Energy, derived from the nitrogen adsorption isotherm using BJH model

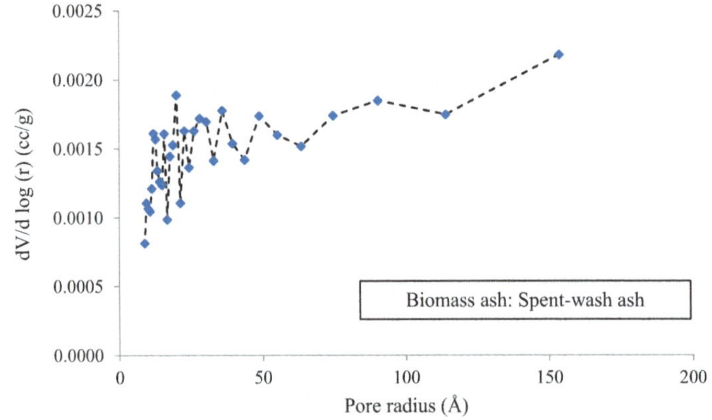

Fig. 2.12 Pore size distribution plot of biomass ash sample generated by combustion of spend-wash from southern part of India, derived from the nitrogen adsorption isotherm using BJH model

sizes to coarser pore sizes. The unilateral trend of increasing pore size distribution from smaller pores to coarser pores indicates that a lot of coarser pores are created during combustion due to rapid escape of volatiles. However, in case of lignites, presence of some finer porous structures within the ashes could cause adsorption of air-entraining agents. Table 2.2 details the results from nitrogen gas adsorption for the studied ash samples.

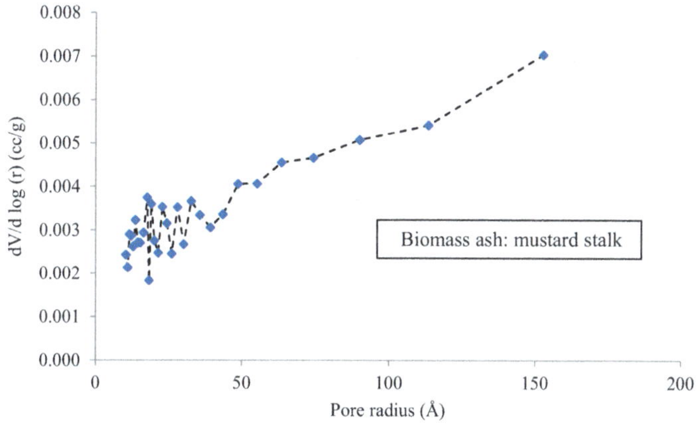

Fig. 2.13 Pore size distribution plot of biomass ash sample generated by combustion of mustard stalk from Rajasthan, India, derived from the nitrogen adsorption isotherm using BJH model

Table 2.2 Pore properties of the coal ash, lignite ash, and biomass ash samples

Samples	BET SSA (m^2/g)	BJH pore volume (cc/g)	Average pore radius (Å)
Fly ash: NTPC-VSTPS	1.27	0.001774	45.86
Fly ash: NTPC-SSTPS	0.68	0.001	27.77
Bottom ash: NTPC-VSTPS	1.81	0.002055	29.25
Bottom ash: NTPC-SSTPS	0.91	0.001816	30.96
Lignite fly ash: JSW Energy	4.24	0.009025	43.67
Lignite bottom ash: JSW Energy	3.72	0.006770	38.88
Biomass ash: spent wash incineration	1.41	0.002172	32.58
Biomass ash: mustard stalk	3.42	0.005463	36.46

2.3.3 Major Oxide Composition of Ash Samples

It is now fairly well established that nearly 316 minerals from 188 mineral groups exist in coals and fly ashes, respectively (Vassilev and Vassileva 2007). Fly ash consists of different components, viz., (a) inorganic constituents made up of non-crystalline matter (such as glassy components), crystalline matter (namely, crystals/grains and multi-mineral aggregates); (b) organic constituents (comprising various degrees of charred materials); and (c) organic minerals (Vassilev 1992, 1994; Querol et al. 1995, 1996; Vassilev and Vassileva 1996, 1997, 2005; Vassilev et al. 2003, 2005). The major and minor elements that occur in both inorganic and organic matter of fly ashes and expressed as oxides are O, Si, Al, Ca, Fe, C, K, Mg, H, Na, Ti, N, P, and Ba, and infrequently Mn, Sr, F, and Cl. The other elements which when detected to be present within fly ashes are mostly trace elements.

Based on their industrial usage, fly ashes are generally categorized into two types: (a) Class F and (b) Class C. While the former type encompasses those fly ashes which are marked by $SiO_2 + Fe_2O_3 + Al_2O_3$ contents ≥70%, for the latter the sum content of the same oxides ranges between 50 and 70% (Vassilev and Vassileva 2007). Class F fly ashes are also marked by low calcium content, which are typically produced by combustion of coals of higher thermal maturity levels or ranks, and typically show pozzolanic properties. Class C fly ashes, on the other hand, are those which are marked by higher calcium contents and produced from low-rank coals, viz., lignites and sub-bituminous coals showing both cementitious and pozzolanic characters.

Table 2.3 presents the major oxide data of few coal and lignite fly ash and bottom ash samples from India. The $SiO_2 + Fe_2O_3 + Al_2O_3$ content of the coal fly ash and bottom ash samples from all sites can be observed to be ≥70%, and consequently falls under Class F fly ash category. Class F ash, in general, are marked by <8% CaO (Vassilev and Vassileva 2007). The lignite fly ash and bottom ash samples from both Jindal South West (JSW; Rajasthan) and Neyveli Lignite Corporation (NLC; Tamil Nadu) are marked by $SiO_2 + Fe_2O_3 + Al_2O_3$ content varying between 62.56 and 66.66 wt%, and consequently falls under Class C ash category. The JSW lignite fly and bottom ash samples are both marked by CaO > 20 wt%, and consequently can be categorized as Class C ash. On the other hand, the NLC lignite fly ash is marked by CaO content of 12.70 wt%, and thus can be categorized as Type CI ash.

Compared to the coal and lignite ash samples, the biomass ash samples showed lower, but variable $SiO_2 + Fe_2O_3 + Al_2O_3$ content varies between 3.06 and 41.60 wt%. Interestingly, the rice husk ash and coconut shell ash had CaO contents lower than 8 wt%, which means the CaO character of these resembles Class F ash. However, as already mentioned, the $SiO_2 + Fe_2O_3 + Al_2O_3$ content of these samples (34.49 and 13.20 wt%) is much lower than the desired $SiO_2 + Fe_2O_3 + Al_2O_3$ content of Class F ash. In contrast to the coal ash samples, Na_2O, K_2O, MgO, and P_2O_5 contents of the biomass ash samples were observed to be much higher than those in coal ash samples.

In addition to the above-mentioned classification scheme, several other ash classification schemes based on their chemistry are available, viz., (a) detrital/authigenic index (DAI; $SiO_2 + Al_2O_3 + K_2O + Na_2O + TiO_2/Fe_2O_3 + CaO + MgO + SO_3 + P_2O_5 + MnO$); (b) SiO_2/Al_2O_3 ratio; (c) free CaO content. Further, fundamental oxide-behavior of coal ash often expressed as base/acid ratio controls a host of factors such as slagging factor, fouling factor, and so on (Vaninetti and Busch 1982). Base/acid ratio is expressed as: $(Fe_2O_3 + CaO + MgO + K_2O + Na_2O)/(SiO_2 + Al_2O_3 + TiO_2)$. The base/acid ratio when divided by Na_2O gives the fouling factor. Other classification schemes are: (a) utilization-based classification based on Si, Al, Fe, Ca, and Mg oxides (Zalkind and Chechik 1971), (b) reactive water-soluble components and amorphous components (Dewey et al. 1996), and (c) petrographic classification scheme based on textural/genetic char features (Hower et al. 2005). However, most of the classification schemes do not take into account the mode of occurrence of the elements in the ash samples.

Table 2.3 Major oxide characteristic of coal fly ash, bottom ash, and biomass ash from several Indian localities

Samples (types/locality)	SiO_2 (wt%)	Fe_2O_3 (wt%)	Al_2O_3 (wt%)	TiO_2 (wt%)	CaO (wt%)	MnO (wt%)	MgO (wt%)	P_2O_5 (wt%)	Na_2O (wt%)	V_2O_5 (wt%)	K_2O (wt%)	SiO_2 + Fe_2O_3 + Al_2O_3 (wt%)
Patratu coal fly ash (NTPC)	55.90	4.09	26.57	1.82	0.75	0.05	0.48	0.36	0.09	0.03	1.01	86.56
VSTPS coal fly ash (NTPC)	54.50	5.20	30.41	1.62	0.59	0.08	0.48	0.17	0.11	0.04	0.71	90.11
VSTPS coal bottom ash (NTPC)	59.75	9.47	20.48	1.35	0.43	0.15	0.38	0.12	0.09	0.02	0.66	89.70
SSTPS coal fly ash (NTPC)	63.10	5.47	22.56	1.50	0.57	0.07	0.48	0.13	0.09	0.02	0.81	91.13
SSTPS coal bottom ash (NTPC)	54.68	18.11	17.58	1.29	0.63	0.26	0.44	0.23	0.08	0.02	0.68	90.37
DSTPS coal fly ash (DVC)	59.61	6.49	22.16	1.49	1.43	0.08	0.98	0.49	0.13	0.03	1.40	88.25
DSTPS coal bottom ash (DVC)	53.66	10.77	19.07	1.22	3.17	0.14	1.15	0.30	0.13	0.02	1.12	83.49
Lignite fly ash (JSW)	37.77	4.65	20.15	2.30	20.75	0.05	1.32	NA	1.00	0.03	NA	62.56
Lignite bottom ash (JSW)	47.31	2.59	16.76	1.76	21.58	0.03	0.61	NA	0.68	0.02	0.02	66.66
Neyveli lignite fly ash (NLC)	37.53	6.61	19.75	1.38	12.70	0.03	3.34	NA	0.37	0.03	NA	63.89

(continued)

Table 2.3 (continued)

Samples (types/locality)	SiO_2 (wt%)	Fe_2O_3 (wt%)	Al_2O_3 (wt%)	TiO_2 (wt%)	CaO (wt%)	MnO (wt%)	MgO (wt%)	P_2O_5 (wt%)	Na_2O (wt%)	V_2O_5 (wt%)	K_2O (wt%)	SiO_2 + Fe_2O_3 + Al_2O_3 (wt%)
Biomass ash (Spice ash)	58.09	0.87	1.44	0.08	1.14	0.05	0.53	0.52	0.20	0.00	1.71	60.40
Biomass ash (Rice husk ash)	22.81	3.79	7.89	0.40	7.38	0.17	6.13	3.94	0.87	0.01	16.11	34.49
Biomass ash (Coconut shell ash)	6.12	5.99	1.08	0.08	2.75	0.60	1.94	2.73	1.71	0.00	18.07	13.20
Biomass ash (Orient Green)	26.45	2.73	5.26	0.48	19.18	0.09	3.16	1.49	0.36	0.01	8.10	34.44
Biomass ash (Kalpatru)	31.78	3.25	6.57	0.57	16.51	0.11	2.68	1.44	0.32	0.01	6.85	41.60
Spent-wash ash (DSCL)	2.51	0.40	0.15	0.01	11.19	0.02	7.18	0.75	0.61	0.00	25.35	3.06
Spent-wash (EID)	14.97	2.32	6.07	0.37	9.40	0.04	6.12	0.80	0.49	0.01	18.45	23.36

Table 2.4 LOI values of coal fly ash, bottom ash, and biomass ash samples from India

Samples (types/locality)	LOI (%)
Patratu coal fly ash (NTPC)	0.90
VSTPF coal fly ash (NTPC)	0.38
VSTPF coal bottom ash (NTPC)	1.92
SSTPS coal fly ash (NTPC)	0.01
SSTPS coal bottom ash (NTPC)	0.34
DSTPS coal fly ash (DVC)	0.01
DSTPS coal bottom ash (DVC)	6.45
Lignite fly ash (JSW)	5.64
Lignite bottom ash (JSW)	4.34
Neyveli lignite fly ash (NLC)	4.11
Biomass ash (Spice ash)	30.34
Biomass ash (Rice husk ash)	23.20
Biomass ash (Coconut shell ash)	52.99
Biomass ash (Orient Green)	23.25
Biomass ash (Kalpatru)	20.60
Spent-wash ash (DSCL)	19.23
Spent-wash (EID)	36.20

2.3.4 Unburnt Carbon

Depending on the combustion efficiency at thermal power plants, a certain fraction of the organic-matter present within the coal may survive after the combustion journey, known as unburnt carbon, and can significantly impact the plans of incorporating ashes for different end usage. As mentioned earlier, the unburnt carbon in ash induces adsorption of AEAs and can be detrimental for concretes. Consequently, a thorough estimate of the percentage of residual carbon in ashes is extremely significant.

For fly ash to be used for concretes, ASTM C618-99 (1999) and ASTM C618-05 (2006) suggest that loss-on-ignition (LOI) should be $\leq$6%. LOI gives an approximation of the content of unburned carbon in ash. However, recent studies have also indicated that LOI measurements can overestimate the remaining carbon content of ashes, due to certain secondary reactions. Earlier studies of Heiri et al. (2001), Fan and Brown (2001), and Styszko-Grochowiak et al. (2004) have shown that reproducibility of LOI data should always be checked, and the presence of volatile organic compounds (VOC) can cause up to 20% exaggerated LOI values. Further, it has also been opined that presence of carbonates can also strongly alter the LOI measurements. Table 2.4 shows the LOI values determined for few Indian ash samples. It can be observed that the LOI values of the biomass ash samples are substantially higher than the other ash samples. Consequently, using biomass ash samples for concrete structures is not suggested, as these may induce adsorption of AEAs.

Table 2.5 Composition of VSTPS fly ash sample corresponding to regions of interest A and B, as shown in Fig. 2.14

Position	Element	Weight (%)
A	C	1.67
	O	42.66
	Fe	35.69
	Al	8.00
	Si	9.63
	Mn	2.35
B	C	3.46
	O	51.17
	Al	20.22
	Si	21.31
	K	1.56
	Fe	2.28

Table 2.6 Composition of SSTPS fly ash sample corresponding to region of interest A, as shown in Fig. 2.15

Position	Element	Weight (%)
A	N	83.30
	O	1.87
	Na	0.45
	Si	0.97
	Sr	4.70
	Zr	8.71

The ash from combustion of coal has been a point of interest for coal petrographers, and consequently, several efforts have been made by some researchers to understand the morphology of unburnt carbon of ashes and categorize them according to their morphology and derivation from parental coal macerals. Readers are referred to the pioneering works of Bailey et al. (1990), Hower et al. (1995), Hower et al. (2005), and Suárez-Ruiz et al. (2007) for further details.

2.4 Micro-morphology of Fly Ash

Scanning electron microscopy (SEM) is widely accepted as one of the best methods to characterize the morphology and identifying the different phases present in ashes. High-resolution images in secondary electron (SE) or back-scattered electron (BSE) modes at high working magnifications along with energy-dispersive X-ray spectroscopy (EDS) have appeared as an excellent tool to characterize ash samples (Vassilev 1992, 1994; Demir et al. 2001; Chen et al. 2004; Vassilev et al. 2004; Kutchko and Kim 2006).

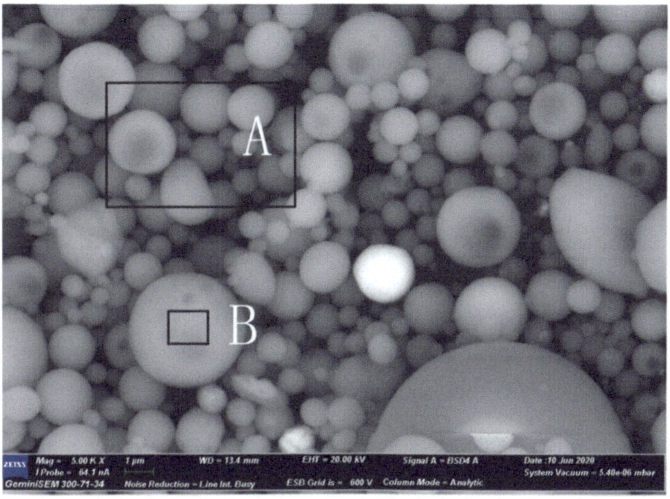

Fig. 2.14 SEM view of a fly ash sample collected from VSTPS thermal power plant, India. A and B represent the positions at which EDS analysis was conducted and presented in Table 2.5

Figure 2.14 shows a SEM view of a fly ash sample collected from VSTPS power plant. It shows the presence of different sized spherical vitreous particles, with smooth texture and appears to be thin walled. The composition of the spherical particles shows the presence of some forms of carbon within the spherical particles, in addition to other elements, viz., oxygen, iron, aluminum, silicon, potassium, and manganese. Presence of such spherical particles, as seen in Fig. 2.14, has been frequently reported by other researchers (Kutchko and Kim 2006). Figure 2.15 shows the presence of similar spherical particles within a fly ash sample collected from a different thermal power plant (SSTPS). In contrast to the fly ash sample from VSTPS, for the fly ash sample from SSTPS, no carbon was detected, but extremely high content of nitrogen was detected through EDS analysis. Additionally, the spheres also showed the presence of zirconium and strontium elements, in addition to silicon, sodium, and oxygen.

In contrast to the fly ash samples, the bottom ash samples showed coarser and irregular-shaped particles, and also greater proportion of unburnt carbon. Further, although some semi-spherical particles were observed to be present within the bottom ash sample, the surfaces were not observed to be smooth as observed for fly ash samples. Rather, they were observed to be marked by the presence of various shaped devolatilization vacuoles and pores (Fig. 2.16 and Table 2.7). The SEM images thus correlate well with the BET SSA and BJH pore volumes, which were observed to be higher for the bottom ash samples relative to the fly ash samples. SEM view of the biomass ash sample also showed the presence of particles with wide range of particles, with varying shapes and morphologies (Fig. 2.17).

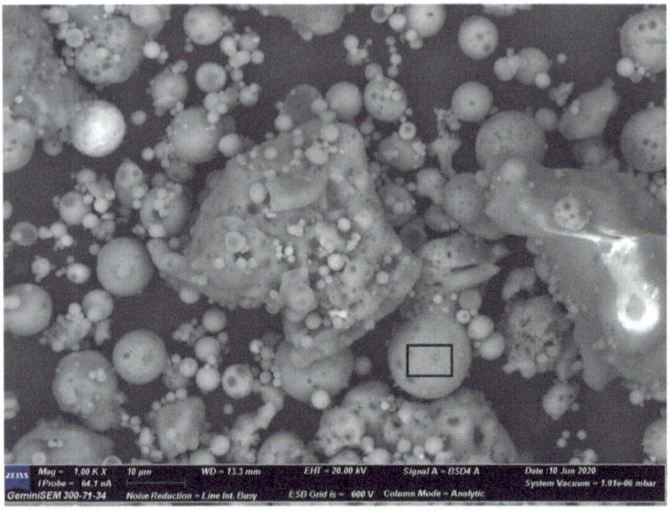

Fig. 2.15 SEM view of a fly ash sample collected from SSTPS thermal power plant, India. A represents the position at which EDS analysis was conducted and presented in Table 2.6

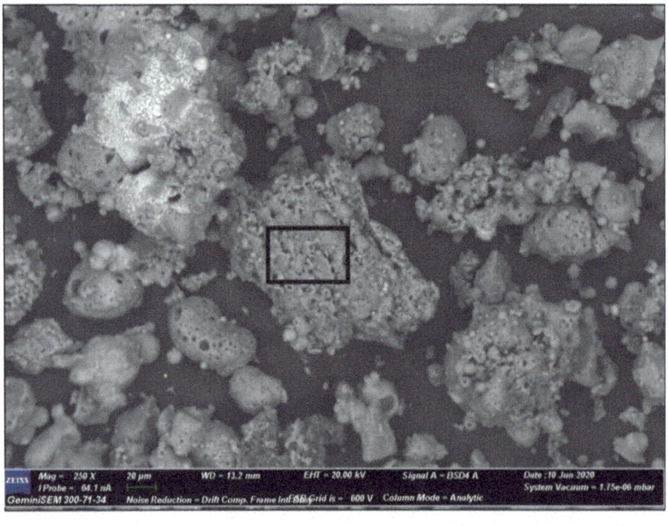

Fig. 2.16 SEM view of a bottom ash sample collected from VSTPS thermal power plant, India. The square block represents the portion at which EDS analysis was conducted and presented in Table 2.6

Table 2.7 Composition of VSTPS bottom ash sample corresponding to region of interest, as shown in Fig. 2.16

Position	Element	Weight (%)
Square	C	11.76
	N	0.84
	O	43.83
	Al	19.04
	Si	20.81
	Ti	0.99
	Fe	2.73

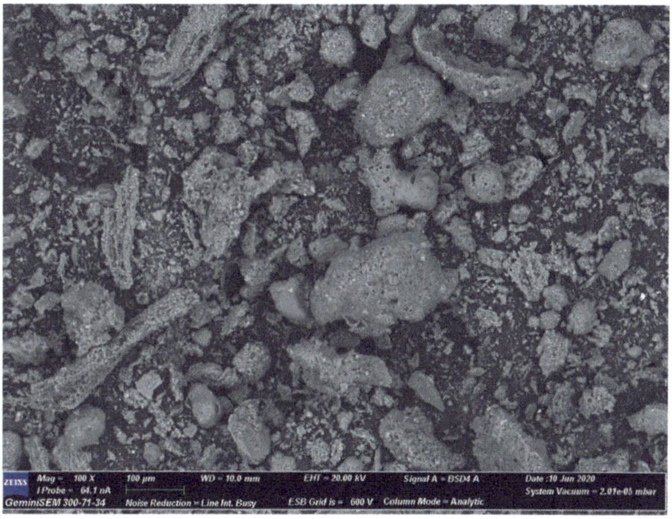

Fig. 2.17 SEM view of a biomass ash sample from India, showing the presence of particles with wide range of sizes and morphologies

References

ASTM (1999) C618-99, Standard specification for coal fly ash and raw or calcined natural pozzolan for use as a mineral admixture in concrete. American Society for Testing and Materials, Philadelphia, USA, 4p

ASTM (2006) C618-05, Standard specification for fly ash and raw or calcined natural pozzolan for use in concrete. American Society for Testing and Materials, Philadelphia, USA, 2p

Bailey JG, Tate A, Diessel CFK, Wall TF (1990) A char morphology system with applications to coal combustion. Fuel 69:225–239

Brown JH (1982) The strength and workability of concrete with PFA substitution. In: Cabrera JG, Cusens AR (eds) International symposium on the use of PFA in concrete. University of Leeds, Department of Civil Engineering, Leeds, UK, pp 151–161

Chen Y, Shah N, Huggins FE, Huffman GP, Linak WP, Miller C (2004) A. Investigation of primary fine particulate matter from coal combustion by computer-controlled scanning electron microscopy. Fuel Process Technol 85:743–761

Demir I, Hughes RE, DeMaris PJ (2001) Formation and use of coal combustion residues from three types of power plants burning illinois coals. Fuel 80:1659–1673
Demirbas A (2001) Biomass resource facilities and biomass conversion processing for fuels and chemicals. Energy Convers Manage 42:1357–1378
Dewey GR, Sutter LL, Sandell JF (1996) Reactivity based approach for classifying fly ash. In: The American power conference, Chicago, Illinois, USA, April, 1996, pp 1–4
Fan M, Brown RC (2001) Comparison of the loss-on-ignition and thermogravimetricanalysis techniques in measuring unburned carbon in coal fly ash. Energy Fuels 15:1414–1417
Hazra B, Wood DA, Vishal V, Varma AK, Sakha D, Singh AK (2018a) Porosity controls and fractal disposition of organic-rich Permian shales using low-pressure adsorption techniques. Fuel 220:837–848
Hazra B, Wood DA, Vishal V, Singh AK (2018b) Pore-characteristics of distinct thermally mature shales: influence of particle sizes on low pressure CO_2 and N_2 adsorption. Energy Fuels 32:8175–8186
Hazra B, Wood DA, Mani D, Singh PK, Singh AK (2019a) Organic and inorganic porosity, and controls of hydrocarbon storage in shales. In: Evaluation of shale source rocks and reservoirs. Petroleum engineering. Springer, Cham
Hazra B, Chandra D, Singh AK, Varma AK, Mani D, Singh PK, Boral P, Buragohain J (2019b) Comparative pore structural attributes and fractal dimensions of Lower Permian organic-matter-bearing sediments of two Indian basins: inferences from nitrogen gas adsorption. Energy Sourc Part A: Recov Util Environ Effects. https://doi.org/10.1080/15567036.2019.1582737
Heiri O, Lotter AF, Lemcke G (2001) Loss on ignition as a method for estimatingorganic and carbonate content in sediments: reproducibility and comparability of results. J Paleolimnol 25:101–110
Hower JC, Rathbone RF, Graham UM, Groppo JG, Brooks SM, Robl TL, Medina SS (1995) Approaches to the petrographic characterization of fly ash. In: Proceedings of 11th international coal testing conference, Lexington, KY, 10–12 May 1995, pp 49–54
Hower JC, Suárez-Ruiz I, Mastalerz M (2005) An approach toward a combined scheme for the petrographic classification of fly ash: revision and clarification. Energy & Fuels 19(2):653–655
Jenkins BM, Baxter LL, Miles TR Jr, Miles TR (1998) Combustion properties of biomass. Fuel Process Technol 54:17–46
Kamruzzaman, M, Islam SMZ, Nizamudoulah, SM, Jingkaojai S (2003) Performance of pulverized fuel ash in cement concrete. In: Proceedings of international workshop and conference on construction management and materials (CONMAT 2003), IIT Kharagpur, Kharagpur, India, 9–11 January 2003, pp 539–548
Karltun E, Saarsalmi A, Ingerslev M, Mandre M, Andersson S, Gaitnieks T, Ozolinčius R, Varnagiryte-Kabasinskiene I (2008) Wood ash recycling—possibilities and risks. In: Röser D, Asikainen A, Raulund-Rasmussen K, Stupak I (eds) Sustainable use of forest biomass for energy. A synthesis with focus on the Baltic and Nordic region. Springer, Heidelberg, pp 79–108
Keefer RF (1993) Coal ashes-industrial wastes or beneficial byproducts? In: Keefer RF, Sajwan KS (eds) Trace elements in coal and coal combustion residues. Lewis Publishers, Ann Arbor, pp 3–9
Kula I, Olgun A, Sevine V, Erdogan Y (2002) An investigation on the use of tincal ore waste, fly ash and coal bottom ash as Portland cement replacement materials. Cem Concr Res 32:227–232
Külaots I, Hurt RH, Suuberg EM (2004) Size distribution of unburned carbon in coal fly ash and its implications. Fuel 83(2):223–230
Kumar R, Kumar S, Mehrotra SP (2007) Towards sustainable solutions for fly ash through mechanical activation. Resour Conserv Recycl 52:157–179
Kuroda M, Watanabe T, Terashi N (1993) Increase of bond strength at interfacial transition zone by the use of fly ash. Cem Concr Res 30:253–358
Kutchko BG, Kim AG (2006) Fly ash characterization by SEM–EDS. Fuel 85(17–18):2537–2544
Mahajan OP (1991) CO_2 surface area of coals: the 25 year paradox. Carbon 29:735–742
McKendry P (2002) Energy production from biomass (part 1): overview of biomass. Biores Technol 83(1):37–46

Potgieter SSV, Potgieter JH, Kruger RA, Spolnik Z, Grieken RV (2005) A characterization of the surface properties of an ultrafine fly ash (UFFA) used in polymer industry. Fuel 84:2295–2300
Psarras P, Holmes R, Vishal V, Wilcox J (2017) Methane and CO_2 adsorption capacities of kerogen in the Eagle Ford shale from molecular simulation. Accounts Chem Res 50(8):1818–1828
Punshon T, Seaman JC, Adriano DC (2002) The effect of flue gas desulfurization residue on corn (*Zea mays* L.) growth and leachate salinity: multiple season data from amended mesocosms. In: Sajwan KS, Alva AK, Keefer RF (eds) Chemistry of trace elements in fly ash. Kluwer Academic/PlenumPress: New York, NY
Querol X, Fernandez-Turiel JL, Lopez-Soler A (1995) Trace elements in coal and their behaviour during combustion in a large power station. Fuel 74:331–343
Querol X, Juan R, Lopez-Soler A, Fernandez-Turiel JL, Ruiz CR (1996) Mobility of trace elements from coal and combustion wastes. Fuel 75:821–838
Ross DJK, Bustin RM (2009) The importance of shale composition and pore structure upon gas storage potential of shale gas reservoirs. Mar Pet Geol 26:916–927
Sajwan KS, Punshon T, Seaman JC (2006) Production of coal combustion products and their potential uses. In: Sajwan KS, Twardowska I, Punshon T, Alva AK (eds) Coal combustion byproducts and environmental issues. Springer, New York, NY
Sarkar A, Rano R, Mishra KK, Sinha IN (2005) Particle size distribution profile of some Indian fly ash—a comparative study to assess their possible uses. Fuel Process Technol 86(11):1221–1238
Sing KS (1985) Reporting physisorption data for gas/solid systems with special reference to the determination of surface area and porosity (Recommendations 1984). Pure Appl Chem 57(4):603–619
Styszko-Grochowiak K, Golas J, Jankowski H, Kozinski S (2004) Characterization of the coal fly ash for the purpose of improvement of indus-trial on-line measurement of unburned carbon content. Fuel 83:1847–1853
Suárez-Ruiz I, García AB (2007) Optical parameters as a tool to study the microstructural evolution of carbonized anthracites during high-temperature treatment. Energy Fuels 21(5):2935–2941
Sun W, Handog Y, Bingem Z (2003) Analysis of mechanism of water reducing effect of fine ground slag, high-calcium fly ash and low-calcium fly ash. Cem Concr Res 33:1125–1199
Thomas Jr J, Damberger HH (1976) Internal surface area, moisture content and porosity of Illinois coals: variations with coal rank. Illinois State Geological Survey Circular 493, 38 p
Tomeczek J, Palugnoik H (2002) Kinetics of mineral matter transformation during coal combustion. Fuel 81(10):1251–1258
Vaninetti GE, Busch CF (1982) Mineral analysis of ash data: a utility perspective. J Coal Qual 1:22–31
Vassilev SV (1992) Phase mineralogy studies of solid waste products from coal burning at some Bulgarian thermoelectric power plants. Fuel 71:625–633
Vassilev SV (1994) Trace elements in solid waste products from coal burning at some Bulgarian thermoelectric power stations. Fuel 73:367–374
Vassilev SV, Vassileva C (1996) Mineralogy of combustion wastes from coal-fired power stations. Fuel Process Technol 47:261–280
Vassilev SV, Vassileva C (1997) Geochemistry of coals, coal ashes and combustion wastes from coal-fired power stations. Fuel Process Technol 51:19–45
Vassilev SV, Vassileva C (2005) Methods for characterization of composition of fly ashes from coal-fired power stations: a critical overview. Energy Fuels 19:1084–1098
Vassilev S, Vassileva C (2007) A new approach for the classification of coal fly ashes based on their origin, composition, properties, and behaviour. Fuel 86:1490–1512
Vassilev S, Baxter D, Andersen L, Vassileva C (2010) An overview of the chemical composition of biomass. Fuel 89:913–933
Vassilev SV, Baxter D, Andersen LK, Vassileva CG (2013) An overview of the composition and application of biomass ash. Part 1. Phase–mineral and chemical composition and classification. Fuel 105:40–76

Vassilev S, Menendez R, Alvarez D, Diaz-Somoano M, Martinez-Tarazona MR (2003) Phase-mineral and chemical composition of coal fly ashes as a basis for their multicomponent utilization. 1. Characterization of feed coals and fly ashes. Fuel 82:1793–1811

Vassilev S, Baxter D, Andersen L, Vassileva C, Morgan T (2012) An overview of the organic and inorganic phase composition of biomass. Fuel 94:1–33

Vassilev SV, Menendez R, Borrego A, Diaz-Somoano M, Martinez-Tarazona MR (2004) Phase-mineral and chemical composition of coal fly ashes as a basis for their multicomponent fly ash utilization. 3. characterization of magnetic and char concentrates. Fuel 83:1563–1583

Vassilev S, Vassileva C, Karayigit A, Bulut Y, Alastuey A, Querol X (2005) Phase-mineral and chemical composition of composite samples from feed coals, bottom ashes and fly ashes at the Soma power station, Turkey. Int J Coal Geol 61:35–63

Vishal V, Ranjith PG, Singh TN (2013) CO_2 permeability of Indian bituminous coals: implications for carbon sequestration. Int J Coal Geol 105:36–47

Vishal V, Singh TN, Ranjith PG (2015) Influence of sorption time in CO_2-ECBM process in Indian coals using coupled numerical simulation. Fuel 139:51–58

Vishal V, Chandra D, Bahadur J, Sen D, Hazra B, Mahanta B, Mani D (2019) Interpreting pore dimensions in gas shales using a combination of SEM imaging, small-angle neutron scattering, and low-pressure gas adsorption. Energy Fuels 33(6):4835–4848

Williams A, Jones JM, Ma L, Pourkashanian M (2012) Pollutants from the combustion of solid biomass fuels. Prog Energ Combust 38:113–137

Zalkind I, Chechik A (1971) Physico-chemical properties of fly ashes and slags and possibilities for their utilization. Energeticheskoe Stroitelstvo 1:18–22 (in Russian)

Chapter 3
Utilization of Coal and Biomass Ash

Abstract Sustainable utilization of the ash generated from the combustion of coal or biomass is a big challenge for the power industry. Huge quantities of ash are generated and, in general, they are disposed-off in ash ponds. However, recent regulatory requirements demand 100% utilization of ash. So many new areas of ash utilization are being explored by the researchers and ash managers. Bulk utilization sectors are cement industry, construction, bricks, landfill, mine back filling, and soil amendment for growing plants. Efforts to enhance the use in value-added low-volume sectors like fertilizer, cenosphere, catalyst support, zeolites, aerogels, and so on are continuously evolving. The heterogeneity of the ash properties is one of the main challenges for advocating a generalized utilization pattern of the ash. Biomass has some typical properties that limit its use for some sectors. However, beneficiation of both coal and biomass ash and use of other additives could improve the suitability of the ashes to multifarious uses.

3.1 Introduction

Coal and biomass ashes have some typical properties that make them qualify for different uses. Though these materials are heterogeneous in terms of physical and chemical properties, their use for different industrial and non-industrial applications has greater potential. National Thermal Power Corporation of India (NTPC) has identified the following major areas of ash utilization (NTPC 2007):

1. Production of Portland-Pozzolana cement and functioning enhancer of ordinary Portland cement (OPC).
2. Construction of ash bricks and other structure commodities.
3. Portion-substitute of OPC in cement concrete.
4. Soil modification for agricultural and wasteland improvement.
5. High-quantity fly ash concrete.
6. Roller-compressed concrete applied for dam and pavement building.
7. Formation of roadblock, structural-fills, low-lying area improvement.

The use of fly ash in different sectors is discussed in the subsequent sections.

A. K. Singh et al., *Ash from Coal and Biomass Combustion*,
https://doi.org/10.1007/978-3-030-56981-5_3

3.2 Agriculture and Forestry

Usage of fly ash as a soil ameliorant has received wide interest during the last four decades. The properties of most of the fly ash favor its suitability for soil application in agriculture and forestry sector. The particle size of the ash in the silt range and the chemical composition, that is, the presence of plant nutrients in ash, makes it a useful material for soil application. There are several research articles and review papers that demonstrate the beneficial role of fly ash on plant development and soil physical–chemical–biological characteristics. Worldwide due to the increase in urbanization and intensive cultivation, the quality of many soils has degraded. Over the years, the crop yields are declining; water and fertilizer use efficiencies are showing a decreasing trend. Many inorganic and organic materials, like lime, gypsum, compost, biochar, farm yard manure, and so on, are used to ameliorate these poor-quality soils. However, many a times, the amendment materials are costly and not economical. They take longer time to enhance the soil aggregation, and to ameliorate other soil structural properties. Coal and biomass ash could be alternative materials to improve the soil quality.

Like coal ash, biomass ash has also some favorable properties that support its application in agriculture and forestry sector. Recycling of biomass ash in agriculture contributes to realize nutrient cycling and decrease the use of commercial fertilizers (Schiemenz et al. 2011b). During plant growth a lot of nutrients taken up by the biomass are stored in the plant parts. After combustion these nutrients are concentrated in the ash. Nutrient-rich biomass ash has direct benefit on improving the soil fertility.

The pozzolanic property of fly ash improves soil aggregation and modifies soil bulk-density, porosity, and water-retention ability. Predominance of silt-sized particles in ash modifies soil texture and infiltration rate. Presence of significant amount of plant nutrients in the ash improves the nutrient pool of the soil. Improvement in the soil physical conditions and availability of plant nutrients helps in the proliferation of soil organisms and the overall soil health is improved by the fly ash. Fly ash prevents leaching loss of plant nutrients. Weathering of fly ash in soils leads to the formation of zeolite layers on ash surfaces, which holds macro- and micro-nutrients, and allows all fly ash particles to host different types of nutrients. The specific benefits of ash on different soil properties and plant growth are discussed in the subsequent sections.

3.2.1 Impact of Ash on Soil Physical Properties of Soil

Different soil characters and their behavior under the influence of fly ash are briefed in Table 3.1. Soil porosity is modified by fly ash application (Campbell et al. 1983). The lighter and smaller hollow ash particles accumulate in the soil voids and modify the soil particle size distribution and other structural properties (Ram and Masto 2010). Both the clayey and sandy textures of soil could be modified by the addition of ash.

Table 3.1 Effect of fly ash on physical properties of soil

	Soil property	Importance	Effect of ash
1	Bulk density	Mass of the soil per unit bulk volume. It is an indication of soil compactness. It influences soil infiltration, root growth, water storage, nutrient solubility, habitation of soil organism, and crop productivity	As the ashes are lighter than the soil, ash application decreases the bulk density. Presence of silt-sized ash particles also decreases soil bulk density
2	Porosity	Space or the voids in between the soil particles and soil aggregates. Helps in water retention, supply of oxygen, and nutrients required for soil organisms and plant roots	Irregular-shaped particles of fly ash aids in loose packing of the soil
3	Water-retention capacity	Quantity of water a soil can retained at field capacity. Soils with high water-holding capacity controls runoff, loss of nutrients, and saves irrigation costs	Fly ash alters soil texture, decreases soil bulk density, and improves soil porosity. Along with these alterations, complex porous structure of the ash helps to increase the water-holding capacity
4	Infiltration rate	Rate of water entry into the soil. Surface soil properties determine the infiltration rate. It is the soils capacity to allow water movement into and through the soil profile. Poor infiltration leads to water stagnation, low aeration, poor soil structure, prone to runoff, and erosion	High proportion of silt-size particles in fly ash promotes infiltration
5	Hydraulic conductivity	Capacity of the soil to transmit water. It affects plant growth by regulating the water distribution in the soil profile	High rate of fly ash use decreases hydraulic conductivity owing to their pozzolanic properties
6	Erodibility	Susceptibility of soil to erosion. Soil erosion leads to loss of surface soil, organic matter, and plant nutrients	Use of fly ash decreases soil erodibility by modifying the permeability characteristics of the soil
7	Permeability	Property of the soil to transmit water and air. It affects the supply of moisture and nutrients in the root zone of plants	Presence of silt-sized particles increases soil permeability

(continued)

Table 3.1 (continued)

	Soil property	Importance	Effect of ash
8	Soil aggregation/structure	Binding of the major soil particles (sand, silt, clay) into soil aggregates. It improves gas exchange, root growth, drainage, and inhabitation of soil organisms	Presence of Ca and pozzolanic properties of the ash improves soil aggregation

The resultant soil texture is generally loamy that is suitable for plant growth (Fail and Wochok 1977). Silt-sized particles of the fly ash decrease soil bulk density and form a favorable soil texture, especially with increased micro-porosity and enhanced water-holding capacity (Ghodrati et al. 1995). The betterment in the soil texture and bulk density improves the workability, infiltration rate, hydraulic conductivity, water-holding capacity, gas exchange, and other physical characteristics of the soil. Fly ash mixing with soils decreases the propensity of the soil to form crusts. Crusts are thin, hard soil surface layers that prevent seedling emergence, water infiltration, and soil aeration. The improvements in the physical properties of soil are widely reported by several researchers. Other than the silt-sized components, the pozzolanic nature of the ash improves the aggregation and the structural properties of the sandy soils. Combined usage of fly ash and organic and inorganic materials decreases the bulk density of the alkaline soil and mine spoil (Jastrow et al. 1979). Wherever the fly ashes are lighter than the soil, the bulk density of the soil decreases after ash addition (Shaheen et al. 2014). Application of high dose of ash significantly improves the water-retention ability of coarse-textured soil, for example, 10% ash addition enhances the water-holding capacity of fine and coarse textured soil by 7.2 and 13.5%, respectively (Campbell et al. 1983). Similarly, several studies support the use of fly ash to bring favorable changes in the physical quality of the soils. Fly ash or bed ash amendment modifies the soil properties from sandy-loam to silty-loam and sandy-clay to loamy (Lu and Zhu 2004). Fly ash addition conserves soil moisture (Seneviratne et al. 2010). Recent works have showed that fly ash addition modifies soil pores network and in turn affects soil water-holding capacity and water evaporation loss. Fly ash increases the water-holding capacity, but the evaporative loss is mainly controlled by the organic matter, so co-application of ash with organic amendment is recommended (Song et al. 2020). Adding lime and fly ash increases the angle of repose and cohesion value. Further, micro-porosity of the soil amended with ash increases and thereby augments its water-retention capacity and decreased the propensity for soil erosion (Yoga et al. 2019). Fly ash further improves the water content of soil accessible to the plant (Adriano et al. 1978; Adriano and Weber 2001). The fine lightweight porous glassy particles of fly ash decrease the bulk density, increase permeability, and optimize the temperature of the soil surface (Lu et al. 2014). High water-retention capacity and the maintenance of soil temperature make the fly ash as a good amendment material for arid soils (He and Shi 2012).

Like coal ash, biomass ash also has small-sized particles that improve the soil physical properties. On contact with water, ash particles swells and clogs the soil pores, affects the soil texture, increases water-holding capacity, decreases aeration, decreases bulk density and hydraulic conductivity, and increases soils resistance against erosion (Bougnom et al. 2011; Etiegni et al. 1991). In a montmorillonite high compaction saline soil, application of bagasse ash improved the mechanical engineering properties of the soil, especially decreased soil compaction and increased hydraulic conductivity (Seleiman and Kheir 2018). Sugarcane bagasse ash can be used to improve the workability of problematic soils; it decreases plastic limit, liquid limit, and plasticity index (Masued 2017). Similarly, earlier attempts were made to use rice husk as a soil stabilizer (Behak 2016).

3.2.2 Impact of Ash on Chemical Properties of Soil

Modification of soil pH and the addition of essential plant nutrients are the major roles of fly ash on soil. The general effects of fly ash on the chemical properties of soil are presented in Table 3.2.

3.2.2.1 Soil pH

Most of the fly ashes are alkaline in nature and are used for ameliorating acid soils. Generally, lime or dolomite is applied for retrieval of acidic soils and mine spoil, but as they are costly, their availability would be limited in the future (Yunusa et al. 2006). Class C fly ash having >15% CaO is widely used to reclaim acid soils. However, Class F fly ash of <15% CaO has limited application to reclaim soil acidity. Class C ash can be substituted for lime to manage acid soils (Parkinson and Coleman 1991). Some silicate minerals like mullite present in the ash scavenges soil hydrogen ions and increases soil pH by formation of silicic acid (Ram and Masto 2014). Alkaline fly ash can substitute about 20% of the $CaCO_3$ for neutralizing acid soil (Phung et al. 1979). Slow weathering of the ash steadily improved the pH of a mine spoil (Seoane and Leiros 2001). Liming effect of fly ash on acidic soils has been widely reported (Lee et al. 2006; Lopareva-Pohu et al. 2011; McCarty et al. 1994; Shaheen and Tsadilas 2010; Tarkalson et al. 2005). Usage of fly ash along with organic materials and lime elevates the pH in many acidic soils (CFRI 2001; Ram et al. 2007; Scotti et al. 1999; Wong and Lai 1996) and coal-mine spoils (Srivastava and Chhonkar 2000). The pH of the soil can increase or decrease depending on the nature of the ash. Fly ashes obtained from high sulfur coal are acidic in nature. Application of acidic fly ashes decreases the soil pH. These ashes are useful for reclamation of alkali soils (Pathan et al. 2003). Appropriate fly ash may be applied depending upon the nature of the soil.

Like coal ash, biomass ash is also used to ameliorate soil acidity due to the presence of significant amount of alkaline elements like K, Ca, Mg, P (Gómez-Rey

Table 3.2 Impact of fly ash on chemical characteristics of soil

	Soil property	Importance	Effect of ash
1	pH	Soil pH measures the hydrogen ion activity. pH affects the availability of plant nutrients, growth and yield of crops, activities of soil biota, etc	The pH of the fly ash is dependent on the nature of the coal burnt. Most of the fly ashes are alkaline. Presence of hydroxide and carbonate salts in ash neutralizes acid soils
2	Electrical conductivity	It is the measure of soluble salt content. High electrical conductivity affects plant growth, and soil–water balances. Crop yield, solubility of plant nutrients, and soil microbial activity were also affected by electrical conductivity	Soluble salts present in the fly ash increases the soil electrical conductivity
3	Nutrient content	Essential plant nutrients like C, H, O, N, P, K, Ca, Mg, Cl, S, Zn, Fe, Cu, Mn, Mo, Ni, B are needed in adequate quantities to complete the lifecycle of plants. Inorganic nutrients should be present in the chemicals forms that are easily absorbed by plant roots	Amelioration of soil pH mediated by fly ash increases the availability of soil nutrients. Further, fly ash contains significant amount of most of these plant nutrients except N
4	Cation exchange capacity	It is an inherent soil property expressed as the ability of the soil to hold cations/nutrients and buffer against soil reaction. This is a measure of the total negative charges on the soil surface	Generally, because of the dilution effect, fly ash decreases cation exchange capacity of the soil
5	Organic C	Soil C is a measure of the organic matter content of soil. It builds soil structure, increases physical stability, improves soil aeration, water retention and movement, decrease soil erodibility, and helps in nutrient cycling	Application of fly ash gradually increases soil organic C due to the increase in plant growth and root biomass

(continued)

Table 3.2 (continued)

	Soil property	Importance	Effect of ash
8	Potentially toxic elements	Soil contamination with potentially toxic elements leads to serious risks to plants, animals, and human beings	Increase in soil pH due to ash application immobilizes the potentially toxic elements into less-soluble forms. However, some fly ashes are enriched with potentially toxic elements that may pose ecological risks. Selection of proper fly ash and other organic amendments to be used along with ash would avoid the soil contamination with potentially toxic elements
9	Organic pollutants	Polycyclic aromatic hydrocarbons (PAHs) and poly chlorinated biphenyls (PCBs) are the two potential organic contaminants from fly ash that could pollute the soils	PAHs content in the ash is directly correlated with the unburnt carbon content of the ash. High pH of the ash could decrease the mobility/availability of PCBs. Proper selection of ash for soil application is essential to avoid potential risks from organic pollutants

et al. 2013; Park et al. 2012). High Ca with low S concentration is another reason for high pH (Vassilev et al. 2013a). Usage of biomass ash, bone-meal, and alkaline slag increased the pH of five Ultisols by 0.63–1.37. Exchangeable soil acidity decreased by 80.1–96.9%; exchangeable potassium (0.4–3.8 times), calcium (1.9–10 times), magnesium (1.7–9.7 times), and available P (0.6–184 times) were also increased significantly (Shi et al. 2016), and the results alike were shown by Shi et al. (2017). Chemical composition and particle size of the biomass ash influence the amelioration of soil acidity. Application of biomass ash to soil increased soil pH and release of nutrients like Ba, Ca, K, Mg, and Na (Cruz et al. 2017). The beneficial role of biomass ash on acid soil is owing to the occurrence of carbonates of calcium, magnesium, and potassium and fine structure (Quirantes et al. 2016). Long-term application of biomass ash progressively increased soil CEC and base saturation (Ca, Mg, K); and concomitantly decreased the exchangeable acidity (Augusto et al. 2008). Change in the soil CEC is due to the modification in soil surface charges mediated by pH (Molina et al. 2007).

3.2.2.2 Salinity

Electrical conductivity, a measure of soluble salts in soils increased with ash application (Matsi and Keramidas 1999; Singh and Agrawal 2010). Decrease in soil electrical conductivity after ash application due to precipitation of salts was also reported (Sinha and Gupta 2005). Bagasse ash decreased soil salinity (up to 30%) and increased soil organic matter and N content (Seleiman and Kheir 2018). Biomass fly ash decreased soil electrical conductivity from 89 to 407 μS/cm (Fernández-Delgado Juárez et al. 2019). Application of high dose of biomass ashes increases soil salinity (Quirantes et al. 2016). The soil salinity will be affected depending on the nature of the ash.

3.2.2.3 Soil Fertility

Fly ash contain many of the elements required for the growth of plant like phosphorous, calcium, magnesium, sulfur, zinc, iron, copper, manganese, molybdenum, and so on. Depending on the chemical association of these elements in ash, the solubility of these elements varies. Many studies were focused on adding other organic and inorganic amends along with fly ash to improve the solubility and plant availability of these elements. Due to the high temperature combustion in the power plants, most of the N originally present in the coal is lost. So, while adding fly ash to soil, application of nitrogenous fertilizer from other sources is required. Fly ash application increases the concentration of P, K, B, Ca, Mg, Mn, Zn, carbonates, bicarbonates, and sulphates in soil (Khan and Singh 2001). Some fly ashes contain significant amounts of S that can ameliorate S deficiency of soils and plant (Chang et al. 1977). Deficiencies of other elements were also corrected by the use of fly ash: Mo (Elseewi et al. 1980), Mg (Hill and Lamp 1980), P/Mn (Carlson and Adriano 1993), B (Martens 1971). Presence of high concentration of B in fly ash is one of the limitation of its use for soil amendment; however, use of weathered ash or pond ash could solve this problem (Page et al. 1979).

The nutrient benefits of the use of fly ash are more pronounced in acidic soil. Normally due to low pH of the acidic soils many of the nutrients are not in available forms. Fly ash increases the pH of acidic soils which in turn increases the nutrient availability. Fly ash application corrected the Zn deficiency of an acidic soil (Schnappinger et al. 1975). Instead, Adriano et al. (1978) reported that fly ash caused Zn deficiency in plants (Adriano et al. 1978). Phosphorus availability is challenging in acidic soils. Adriano et al. (1980) reported that the P availability was increased by fly ash application. P-fixing capacity of acid soils was decreased significantly by fly ash addition (Masto et al. 2013). The effects were synergistic when fly ash was added along with organic amendments (Palumbo et al. 2004). Many recent studies also support the use of fly ash for ameliorating nutrient arability in acid soils (Harper and Mbakwe 2020; Rahmawati et al. 2020; Rout et al. 2016).

Solid biofuel ash contains significant amount of plant nutrient (Ca, K, Mg, P, B, Cu, Mn, Mo, Zn) (Demeyer et al. 2001; Biedermann and Obernberger 2005; James et al. 2012; Pedišius et al. 2019). Dry olive-cake ash (P, K, and Cu) and wood ash (Zn) significantly increased the available fraction of soil nutrients (Quirantes et al. 2016). Application of 5% wood ash increased the plant nutrient content in soil, whereas 1% ash did not show any significant effect (Ochecova et al. 2016). Compared to wood ash, straw ash maintained high concentration of Ca, K, Mg, and P in soil solution.

Supply of high amount of K through straw ash releases Ca and Mg from soil sorption sites (Mercl et al. 2016).

3.2.2.4 Potentially Toxic Substances

Presence of some potentially toxic elements in fly ash can hinder plant growth and soil biological activity. Application of high dose of ash increased the concentration of As, Se, Mo in some herbs used as livestock feed (Tolle et al. 1983). Fly ashes also contain some radioactive elements. Mittra et al. (2005) reported elevated radioactivity of ^{226}Ra, ^{228}Ac, and ^{40}K in soil treated with fly ash. However, Goyal et al. (2002) found the radioactivity of soil mixed with fly ash was within the permissible limits. Immobilization of heavy metals by fly ash in mining areas and acid mine drainage have been reported by many studies (Ram and Masto 2010). Combined usage of fly ash and peat decreased the availability of Cu and Pb in soil (Kumpiene et al. 2007). A meta-analysis of data revealed that 0–5% ash application increased the concentration of plant nutrient and some potentially toxic elements (As, Cd, Cr, and Se) in plant (Yu et al. 2019).

Biomass ash may contain significant amount of potentially toxic elements, polycyclic aromatic hydrocarbons (PAH), and polychlorinated biphenyls (PCB) (Bundt et al. 2001). Effect of biomass ash on agricultural soils is scarce, but there are some studies on forest soils. Most of the studies could not find any significant accumulation of potentially toxic elements on soil. Most of the toxic elements were bound to soil organic matter in non-available forms (Augusto et al. 2008). Ash application increased the soil pH which consequently decreased the metal availability (Dimitriou et al. 2006). The alkaline nature of the ash decreases the mobility of toxic elements from soil to plant (Ochecova et al. 2014).

3.2.3 Impact of Ash on Biological Properties of Soil

The data on the impact of fly ash on soil biological properties are limited. Fly ash can influence the soil microorganisms, fauna and other macroorganism directly due to its effect on soil pH, salinity, presence of potentially toxic elements, essential nutrients, and indirectly due to the modifications of physical properties of soil (Carlson and Adriano 1993). Pandey and Singh (2010) found that the soil microbial respiration, N mineralization, and enzyme activity were affected by fly ash. In a sandy soil, fly ash application decreased the microbial respiration (Wong and Wong 1986). Presence of potentially toxic elements and modification of soil pH can affect the microbial growth (e.g. Rhizobacteria) as these microbial community has very specific tolerance level of soil pH and metal contents (Pandey and Singh 2010). In an acidic Alfisol, Lal et al. (1996) reported that the microbial activity increased up to 8% ash application, and higher activity was observed for combined application of farm manure along with ash, whereas Pichtel and Hayes (1990) found that high-rated fly ash adversely affected the cycling of nutrient and microbial decomposition processes in soil. Soil enzyme activities increased up to 10 t/ha ash application and thereafter decreased (Sarangi et al. 2001). Application of lignite ash affected the growth of pathogenic soil microorganisms (Karpagavalli and Ramabadran 1997). There are few studies that showed the beneficial effect of fly ash on soil organisms: Rhizobium sp. and P-solubilizing bacteria (Sen 1997), Pseudomonas striata (Gaind and Gaur 2002), Salmonella, and total coliforms (Wong et al. 2001), arbuscular mycorrhizal fungi and gram-negative bacteria (Schutter and Fuhrmann 2001), diazotrophs and phosphobacteria (Gaind and Gaur 2004), bacteria, and actinomycetes (Vallini et al. 1999; Yeledhalli et al. 2010). Some studies identified negative role of fly ash on bacterial and fungal count of soils (Kucharski et al. 2006).

Microbial activities in soil are generally determined through the activities of soil enzymes. Dehydrogenase activity is an endogenous enzyme responsible for microbial oxidation of soil organic matter, whereas the hydrolases activities play a major role in nutrient cycling and are responsible for transformation of different organic substrates. For example, phosphatase enzyme converts the organic P compounds into available P. High pH and presence of soluble salts could be a significant factor of fly ash to suppress the activities of soil enzymes (Elliott et al. 1982). Soil phosphatase, sulfatase, dehydrogenase, and invertase were held-back by fly ash application, however, catalase activity was not influenced by fly ash (Pichtel and Hayes 1990). Earthworm activity, soil respiration, and enzymes were not affected by fly ash addition up to 2.5%. Studies on the microbial activity on ash amended soils are very few, however, there are very few studies supporting the presence of microbial activity in fly ash dumps colonized by different plant species (Kumar et al. 2008; Rau et al. 2009). Overall, improvement in the physical and chemical properties of soil brought by fly ash provides a better environment for the establishment and the growth of the soil biological system (Pandey and Singh 2010; Ram and Masto 2010, 2014; Ram et al. 2006, 2007). Combined application of fly ash and biochar increased the population of bacterial, actinomycetes and fungal, which is due to the overall enhancement of soil characters (pH, nutrients) (Munda et al. 2016).

Application of biomass ash to soil increased the soil microbial activity and species composition (Augusto et al. 2008). Nitrogen cycling microorganisms were stimulated by ash application (Ozolincius et al. 2006). Increase in soil pH and dissolved organic C favorably stimulated the microbial activity (Bååth and Arnebrant 1994; Jokinen et al. 2006). Soil decomposing animals like enchytraeids and micro-arthropods were not influenced by ash application (Liiri et al. 2002; Lundkvist 1998). Soil organic matter mineralization was increased by ash (Khanna et al. 1994).

3.2.4 Impact of Fly Ash on Plant-Growth and Plant-Yield

Usage of fly ash augmented the yield of several crops: rice (Dwivedi et al. 2007; Lee et al. 2006; Munda et al. 2016; Sarangi et al. 2001; Singh and Siddiqui 2003), cabbage (Kim et al. 1997), rye grass (Matsi and Keramidas 1999), Triticum vulgare (Tsadilas et al. 2009), C. cajan (Pandey et al. 2009a), maize, eggplant (Tripathi et al. 2009), Vigna radiata (Singh and Agrawal 2010), Mentha piperita (Kumar and Patra 2012), and so on. The percent increase in plant yield due to ash application is presented in Fig. 3.1. Enhancement in the physical, chemical, and biological characteristics of soil helps in boosting the growth and yield. However, the effect of ash on crop performance depends on the nature of the ash and the application dose. In contrast, Singh et al. (2008) reported significant decrease in the yield of *Beta vulgaris* due to fly ash application. Sikka and Kansal (1995) found that lower rate (2–4%) of fly ash application increased rice yield, whereas higher rate (8%) decreased the yield. Co-usage of fly ash with other organic substances significantly increased the plant yield by favorably modifying the soil pH, modulating the nutrient availability and immobilization of some potentially toxic elements (Pandey 2020a).

A compilation of large data showed that the increase in plant-biomass yield was affected by the ash application dose. At 0–5% ash application, the yield increase was 11.6%; for 5–10% ash dose was 22.1%; for 10–25% ash, it was 29.2%. Higher dose of ash (50–10%) decreased the plant biomass yield by 45.8% (Yu et al. 2019). Reduction in crop yield is probably due to the presence of high concentration on some potentially toxic elements in unweathered fly ash (Sharma and Kalra 2006).

Many naturally growing tree species and plant species were found to be growing well on the ash-filled basins and low-lying lands. Naturally growing species can adopt and easily survive against the stress conditions (Pandey and Singh 2010). These species can tolerate toxicity as well as deficiency of nutrients (Gajić et al. 2016; Pandey et al. 2012a, b). Some of the prominent naturally growing tree species on fly ash sites are Acacia nilotica, Cassia tora, and Prosopis juliflora (Pandey 2020b). Re-vegetation of ash dumps are generally accompanied by application of organic amendments, microbial inoculants, and bio-fertilizers.

Biomass ash also increased the crop yields. In Cambisol, straw ash or wood ash did not increase the wheat grain yield, but in Fluvisol, both types of ash increased the grain yield: 115% for straw ash; 120% for wood ash (Mercl et al. 2016). Many studies

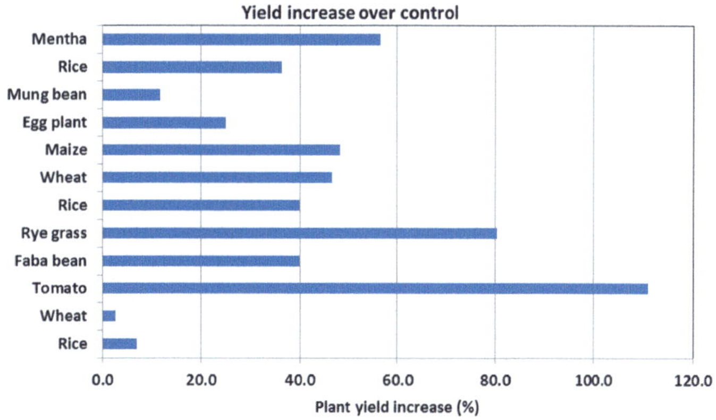

Fig. 3.1 Effect of fly ash on the yield of different plants

reported increase in crop yield due to biomass ash application: 25–69% for winter wheat (Etiegni et al. 1990); 18–25% for barley (Patterson et al. 2004); 7–23% for spring wheat (Ochecova et al. 2014); and 16% for barley (Schiemenz et al. 2011a). Studies show that the supply of Ca, K and the increased use efficiency of P nutrients are the major reasons for increase in crop yield. In a saline soil, bagasse ash increased wheat grain yield by 6.50% (Seleiman and Kheir 2018). Application of palm bunch ash along with mineral fertilizers increased the yield of maize (Adjei-Nsiah 2012). Biomass ash is mostly applied in forest soils. Effect of ash on improvement in tree growth was observed only for organic soils, but not on mineral soils. This is due to the lack of N content in the ash; and the inherent N supply is limited in mineral soils (Augusto et al. 2008).

Though many studies support the application of coal and biomass ash for soil application and for growing plants, some reports alerted the possibility of carryover of potential toxic elements. In conclusion, fly ash use should be specific depending on the quality of the ash and soil and climatic conditions (Bi et al. 2003; Jambhulkar et al. 2018). Application of fly ash in agriculture has been recommended worldwide (Basu et al. 2009; Manoharan et al. 2010; Pandey and Singh 2010; Ram and Masto 2014; Ram et al. 2006, 2007; Shaheen et al. 2014). Carryover of potentially toxic elements has been reported by some studies (Manoharan et al. 2010; Bhattacharya et al. 2012; Pandey et al. 2009b). Thus careful thought should be given on the selection of the nature of the ash and the soils to be applied.

3.3 Utilization of Fly Ash in Cement and Concrete

Cement is used as a binder in construction that adheres to other components and sets into a hard mass. Cement concrete consists of cement coarse aggregates (gravel), fine aggregates (sand), and water. In typical concretes, due to reaction between cement and added water, a gel-like mass is formed that helps in binding different components, and ends up forming a hard material on solidification. Fly ash has been used as a pozzolana in concrete for more than 100 years. Pozzolanic fly ash, particularly the silica and aluminum silicates phase of fly ash chemically reacts with lime to produce cementitious material.

Fly ash has more amorphous phase than the bottom ash. Though both fly ash and bottom ash have comparable chemical composition, the bottom ash contains more of non-reactive crystalline silica which shows less pozzolanic properties. Moreover, the bottom ash is coarser due to the fusion of particles during cooling. Hence, bottom ash is not generally used in cement concrete. Fly ash could be directly added to concrete as one of the components of the mixer or as a constituent of blended cement. It is used in a variety of cement-based products like concrete, concrete block, and brick. It does not have cementitious property, but the Si and Al oxides present in the ash when reacts with $Ca\,(OH)_2$ and water produces compound with cementitious properties. The major reaction product is calcium silicate hydrate. Class C fly ash has significant amount of Ca that shows both pozzolanic and cementitious property. These ashes exhibit self-hardening properties; when react with moisture it forms ettringite, monosulfoaluminate, and calcium silicate hydrate. Similar types of compounds are formed when cement reacts with water, but the reaction rate is slower for fly ash and takes longer time (90 days) to gain strength for the fly ash added concrete.

Portland-Pozzolana cement (PPC) is manufactured by inter-grinding Portland cement clinkers with fly ash or by thoroughly and consistently blending Portland cement and fine-fly ash. Grinding is preferred more. When blending every care should be taken that the components are intimately mixed. Generally, 15–25% fly ash is thoroughly blended, grinded with 65–75% Portland clinker, and 3–6% gypsum. Due to the presence of significant amount of Al_2O_3 and SiO_2 in fly ash, it is used for substituting a portion of other raw materials (clay, shale, etc.) used for manufacturing of cement. Sometimes, the Fe_2O_3 in fly ash is also used to replace iron-based raw material for cement production. As per the Indian Standards, the chemical and physical characteristics required for fly ash to be used in PPC are given in Table 3.3. PPC produces less heat, offers great resistance to chemically reactive water, and decreases the liberation of $Ca(OH)_2$ during setting and hydration. It is further effective for construction in marine and hydraulic conditions. It is equivalent to the grade 33 Portland cement. However, this strength is achieved at a later stage.

Coal ash is also used as an admixture in cement mortar and concrete. Fly ash, bottom ash, pond ash, or mound ash either raw or in processed form could be used as admixture. They are introduced to the batch either before or during mixing to modify freshly mixed, setting, or hardened characteristics. As per Indian Standards (IS 2013), the requirement of the ash to be used as admixture is almost similar to that of Table 3.3, except SO_3 is allowed up to 5%, loss on ignition up to 7.0%, particle size up to 50% (retained on 45 μ sieve), and minimum specific surface area of 200 m^2/kg.

It is now increasingly understood that the strength and durability of concrete in different environment does not increase with increase in proportion of cement in concrete. Supplementary material like fly ash is used in concrete to increase the strength and durability. Concrete is produced by mixing ingredients like cement, coarse aggregates, fine aggregates, water, and fly ash. Cement reacts with water and forms a gel-like material which binds the other components of the concrete into a stone-like hard mass. Use of more cement will liberate more heat and may disturb the concrete. Further, excess lime liberated during the hydration reaction of cement gets solubilized and increases the permeability of the concrete which weakens the durability of the concrete. If fly ash is present in the concrete mix, the excess lime released will react with fly ash (Eq. 3.1) and form additional cementitious materials. This reaction will continue till lime is existent in the concrete. Fly ash may be introduced into concrete either as blended cement or as separate components during concrete mixing. Presence of alumina and silica in fine-sized ash particles react with calcium hydroxide to form cementitious materials.

$$H_4SiO_4 + Ca(OH)_2 \rightarrow CaH_2SiO_4 \cdot 2H_2O \quad (3.1)$$

The amorphous silica reacts with lime and water to form calcium silicate hydrate. Fly ashes with high Ca content show both pozzolanic and hydraulic activity. A

Table 3.3 Requirement of fly ash physical and chemical properties for use as pozzolana for production of cement and for part-substitution of cement in cement mortar and concrete

				Indian standard	
S. No.	Properties	Unit		Siliceous fly ash	Calcareous fly ash
1	$SiO_2 + Al_2O_3 + Fe_2O_3$	Min	%	70	50
2	SiO_2	Min	%	35	25
	Reactive SiO_2	Min	%	20	20
3	MgO	Max	%	5.0	5.0
4	SO_3	Max	%	3.0	3.0
5	Na_2O equivalent available alkalis	Max	%	1.5	1.5
6	Total chlorides	Max	%	0.05	0.05
7	Loss on ignition	Max	%	5.0	5.0
8	Specific surface area	Min	m^2/kg	320	320
9	Particles retained on 45 μ sieve	Max	%	34	34
10	Lime reactivity (average compressive strength)	Min	N/mm^2	4.5	4.5
11	Compressive strength at 28 days	Min	N/mm^2	Not less than 80% of the strength of corresponding plain cement mortar cubes	
12	Soundness by autoclave test (expansion of specimen)	Max	%	0.8	0.8

maximum of 35% fly ash could be applied in concrete mixture. The optimum dose of fly ash fluctuates with the nature of the ash and the indented properties of the concrete. The quality of the fly ash concrete is dependent on the physical and chemical nature of the ash. The use of fly ash in concrete decreases the uses of cement in concrete, thus this substitution will have economic and environmental benefits.

3.3.1 High Volume Fly Ash Concrete

Canada Centre for Mineral and Energy Technology (CANMET) introduced the high-volume fly ash concrete (HVFA) in 1985. According to Malhotra and Mehta (2002), HVFA should have minimum 50% fly ash in the cementitious material; low water content (<130 kg/m^3); and less cement (<200 kg/m^3) is desirable. For better performance of the HVFA, use of super-plasticizers is recommended especially due to the very low water to cement + fly ash ratio. Addition of nano SiO_2, silica fume, polyester fiber, and chemical activators (Na_2SO_4, $CaCl_2$, NaOH) improves the compressive strength and other properties of HVFA (Rashad 2015). One of the major disadvantages of HVFA is reduction in mechanical strength and abrasion resistance. However, HVFA has some specific advantages than the conventional concrete on flow-ability, pump-ability, compatibility, better surface finish, slower setting time, superior dimensional stability and resistance to cracking, resistance to chloride ion penetration, cost economy, and environmental friendliness (Mehta 2004). HVFA is generally used in construction of large structures like dams, foundation and retaining walls, marine slipways, sewage treatment works, concrete viaducts, and so on (Marinković and Dragaš 2018). Deo (2015) reported that the improvement in the workability of the HVFA (50% ash) is due to the 'ball bearing' action and the presence of spherical ash particles. Rivera et al. (2015) reported that the chloride ion permeability of HVFA is significantly less than the OPC concrete. Pozzolanic reaction between fly ash and calcium hydroxide is responsible for the improvement in the mechanical properties and durability of HVFA concrete (Deschner et al. 2012). However, compressive strength of HVFA concrete is much lower than the conventional concrete (Sun et al. 2019).

3.3.2 Roller-Compacted Concrete

In roller-compacted concrete fly ash is applied for partial substitution of ordinary Portland cement. This type of concrete is generally used for dam construction due to less heat produced by fly ash hydration compared to the ordinary concrete. Thus, the possibility for crack development is decreased. Due to less use of water in roller-compacted concrete, the mix is drier and has no slump. This concrete is transported, spread, and compacted on the targeted site by mechanical means using dumper, dozers, and rollers. Roller-compacted concrete is positioned in a mode similar to that of paving; the item is transported by dump trucks or conveyors, distributed by bulldozers or specially tailored asphalt pavers, followed by compacting it using vibratory-rollers. It is also used in road construction as this concrete is easy and fast to construct and requires less machineries than the asphalt road construction (Debbarma et al. 2020). There are reports suggesting 70% fly ash replacement for cement in roller-compacted concrete. The other areas for use of roller-compacted concrete are large parking lots, storage yards, staging areas, street pavements, low speed roods, and for high speed roads (for the foundation/sub-base course). Use of fly-ash-based roller-compacted concrete has been studied and demonstrated extensively (Adamu et al. 2018; Lam et al. 2018; Mardani-Aghabaglou and Ramyar 2013; Wang et al. 2018a).

3.3.3 *Self-compacting Concrete*

These types of concretes are characterized by flowage properties, which becomes compressed under its own weight without segregation (Sua-Iam and Makul 2015). It has much higher flow-ability especially to be used in heavily congested reinforcement and the material is homogenous. Fly ash is used as a raw material in self-compacting concrete (up to 60% cement replacement) (Bouzoubaâ and Lachemi 2001). Though this concrete has low early strength, the properties are comparable with conventional concrete after 28 days curing. However, increasing fly ash proportion beyond a certain limit leads to significant reduction in strength (Khatib 2008; Singh et al. 2019) and may require excessive use of super-plasticizers (Singh et al. 2019). Bottom ash could also be used as a binder to replace cement or as an aggregate or sand replacement (Sua-Iam and Makul 2015). Self-compacting concretes with high proportions of fly ash have practical use in civil engineering jobs where performance is not dependent on the primitive compressive strength.

3.3.4 *Use of Biomass Ash in Concrete*

Use of biomass ash in concrete is not recommended as per the ASTM guidelines (ASTM C 618). Presence of high amount of alkalies, sulfates, and chlorides in biomass ash prevents the use of biomass ash in concrete. However, there are few studies that support that the blended ash (coal ash + wood ash) or the ash generated from co-firing of coal and biomass could be used in concrete with comparable performance to the use of cement without ash substitution (Christopher and Kimberly 2015; Wang et al. 2008).

Biomass ash has more alkalies (K, Ca, Na) and less Si and Al than the coal ash. Further, this composition varies widely with the nature of the biomass fuel feed. However, some biomass ash, like paddy husk, wood, or sugarcane bagasse ash, has comparable pozzolanic properties as coal ash. Biomass ash blended with Class C and Class F fly ash showed similar functioning to fly ash relating to long-standing concrete compressive strength. Among the different biomass ashes, rice husk and sugarcane bagasse ashes are the two main ashes that have been studied extensively for use in cement and concrete.

The chemical composition of rice husk and sugarcane bagasse ash is presented in Table 3.4. Studies showed that use of ground rice husk ash produces concrete with better properties. Rice husk ash with fine particles produces robust and thick mortar due to better dispersion and filling result. However, grinding of rice husk ash is not required as this ash has high internal surfaces that helps in attaining pozzolanic activity (Mehta and Monteiro 2006). Rukzon et al. (2009) opined that the mortar displays enhancement in pozzolanic reactions, in contrast to that of regular concretes. Self-consolidating concrete prepared by mixing 0, 5, 10, 15, 20, 25, and 30% of rice husk ash was evaluated by Kannan and Ganesan (2014). The compressive strength of

Table 3.4 Ash composition of rice husk and sugarcane bagasse ash (*Source* Sua-Iam and Makul (2015))

	Composition (%)	Rice husk ash	Sugarcane bagasse ash
1	SiO_2	82.60–93.40	56.40–78.30
2	Al_2O_3	0.21–0.40	6.74–9.70
3	Fe_2O_3	0.10–0.50	3.61–5.77
4	CaO	0.48–0.90	2.15–8.40
5	MgO	0.20–0.70	1.10–2.97
6	SO_3	0.16–0.40	0.21–2.87
7	LOI	1.27–11.90	0.42–15.34
8	$SiO_2 + Al_2O_3 + Fe_2O_3$	83.50–94.70	75.0–90.5

the concrete with 15% ash was higher than that of the normal concrete. Permeability properties were also better for 15% rice husk mix. Rice husk ash was also used to replace fine aggregates in the concrete mix; however, this increased the water required in the mix owing to the enhanced volume fraction and surface area of binder in the presence of ash. The unit weight of the self-consolidating concrete decreased with increasing ash content, and the compressive strength decreased with increased ash content (Sua-iam and Makul 2013b). Published literature shows that the use of rice husk ash improved compressive and flexural strengths, improved durability, reduced the effects of alkali-silica reaction, reduced shrinkage of concrete, increased resistance to acid attack and chloride penetration, increased water requirement but decreased setting time (Singh 2018).

Presence of high amount of amorphous silica in sugarcane bagasse ash makes it pozzolanic. This property can be exploited to use bagasse ash for partial replacement of clinkers in cement manufacturing (Fairbairn et al. 2010). It can also be used for partial replacement of cement in concrete (Loh et al. 2013). The pozzolanic activity of the bagasse ash depends on the size and fineness of its particles (Cordeiro et al. 2008). Presence of high amount of unburnt carbon is one of the limitations for use of sugarcane ash in cement or concrete (Table 3.4). Blended cement concoctions arranged by adding Portland cement with bagasse ash showed that the setting times increased than the normal cement and the percentage of free lime diminished with rise in quantity of bagasse ash. Compressive strength of mortar is negatively related to the particle size of the bagasse ash; mortar prepared with finest size ash has higher compressive strength and added pozzolanic behavior (Cordeiro et al. 2008). Amin (2011) showed that up to 20% bagasse ash can be possibly blended with Portland cement provided that the ash is well-burnt. These blends showed high early strength, decreased water permeability, and defiance to chloride permeation and diffusion. Sugarcane bagasse ash was also utilized for replacing fine aggregate in the concrete mix (Sua-iam and Makul 2013a). Compared to normal concrete, the ash-substituted (for fine aggregate) concrete presented higher water which in turn led to decreased compressive strength. Overall, processing of bagasse ash, like grinding and calcination, improved the property of the ash and subsequently the quality of cement or concrete (Payá et al. 2018).

Wood ash has high LOI and alkali content with very low SiO_2. Application of wood fly ash for partial sand substitution yielded satisfactory low-quality concrete (Berra et al. 2015). Palm oil fuel ash has high amounts of amorphous Si and Al oxides and it could be considered as a pozzolanic material. Use of ground biomass ash could modify the pore structures of the concrete and favorably alter chloride transport in concrete (Demis et al. 2014). Bottom ash of the olive plants trimmings could be employed as a replacement for Portland cement (Sklivaniti et al. 2017). Though the properties of most of the biomass ashes are not favorable for use in concrete (Berra et al. 2015), grinding and washing of biomass ash could improve its quality for use in cement and concrete.

3.3.5 Advantage of Using Fly Ash in Concrete

1. Comparatively lower amount of water is required in fly ash added concrete than the cement concrete for similar workability.
2. Better workability than the Portland cement concrete. The spherical-shaped particles of the ash decrease the friction and increase the flowability of the concrete with ease to pump. The other factor is the increase in the volume due to the dispersion caused by the presence of finer fly ash particles of same surface charge.
3. Decrease in water requirement decreases the bleeding of the fly ash concrete.
4. Fly ash decreases the rate of heat produced by the concrete due to the slow reactions. This reduced heat decreases the thermal cracks and improves soundness of the concrete.
5. Though the early-age strength of the fly ash concrete is lower, the long-term strength (56 days) after proper curing is higher for fly ash concrete.
6. Long-term flexural and tensile strength of fly ash concrete increases due to the continuous pozzolanic reaction between fly ash and the excess lime released from cement.
7. Decreased creep after long-term curing.
8. Low drying and shrinkage than the conventional concrete.
9. Decrease in permeability of the concrete due to the better filling of the fine-sized ash particles in the concrete matrix.
10. Increased resistance to sulfate attack and alkali-silica reaction due to decrease in content of lime and alkalis available for such reactions.
11. Improves resistance to carbonation due to the decrease in permeability.
12. Decreases corrosion of steel in the concrete due to decreased permeability which arrests the ingression of chlorides and carbonization reaction.
13. Overall, the amount of cement required in the concrete is decreased for providing the same level of strength.

3.3.6 Disadvantages of Using Fly Ash in Concrete

1. The requirement of air entrainment admixtures increases with the use of low Ca and high carbon fly ashes.
2. Low Ca fly ashes increase the time of initial and final setting of concrete.
3. Increase in settling time will take longer finishing and curing time which delays the further operation.
4. Due to the lower strength during the early ages, freezing and thawing may affect the quality of the concrete.

Though addition of fly ash to concrete has several advantages (decreases heat of hydration, permeability, and alkali aggregate reaction; improves workability; increased resistance to sulphate attack and corrosion) (NTPC 2007), the heterogeneity of fly ashes poses a major problem to advocate a generalized recommendation of fly ash use in concrete.

3.4 Bricks, Blocks, and Lightweight Aggregates

3.4.1 Bricks

Lime silicate bricks were originally made in 1898 using lime and sand followed by steam curing. Use of fine sand in this process initiated the thought process to replace fine sand with fly ash. Owing to pozzolanic properties, fly ash reacts with lime to form high strength materials of superior quality. Brick making from clay is a successful process. There are about 50,000 brick-making units in India producing more than 100 billion bricks every year. These bricks are meeting the local needs for the construction industry. However, the clay bricks have low strength, poor technical features related to sound and thermal insulation, high requirement of mortar during construction, inconsistent quality and appearance, and so on.

Summary of different brick making technologies is presented in Table 3.5. Fly ash bricks consist of fly ash, fine aggregates (river sand, stone dust), cement or lime, gypsum, and water mixed in appropriate proportions, compressed (150–200 kg/cm^2), and cured in water or steam. Both Class F and Class C fly ashes are used for manufacturing of ash bricks. Class C ash has self-cementing properties. These bricks can be prepared either by using cement or lime. Fly ash bricks have many advantages than the conventional clay bricks. Ash brick can be produced with or without frog; uniform size; less mortar/plaster required during construction. Manufacturing of ash bricks requires less energy than the conventional clay bricks and can be fabricated even at the construction sites with minimum machineries. A hydraulic press is required for making these bricks. Appropriate proportion of ash, sand, lime/cement, and water are mixed and placed in the molds of the press. After pressing initially, the bricks are air cured for 2 days followed by wet curing (water spraying) for 14 days. The

Table 3.5 Different processes for making bricks from fly ash (*Source* NTPC (2020))

	Technology	Process involved
1	Burnt clay fly ash bricks	Appropriate quantity of fly ash is mixed with soil. Brick molds are produced from the homogenous mixture of soil, fly ash along with required amount of water. Molds are air-dried and fired. Fly ash, bottom ash, or pond ash could be used in this process. Generally, 40% fly ash could be added; however, if the clay content of the soil is high, up to 60% fly ash could be added. Fuel requirement could be decreased by mixing rice husk or rice husk ash containing high unburnt carbon
2	Flux-bonded bricks	Fly ash (80%), good quality clay, a flux-making additive, and required water are mixed and a mold is prepared using a press. Then fired at 800–1000 °C. Apart from bricks, roof tiles, floor tiles/blocks, paving tiles of desired porosity could be produced by this technology
3	Fly ash lime gypsum bricks	Fly ash (50–65%), sand/stone dust (20–30%), hydrated lime (8–12%), and gypsum (5%) are mixed with required quantity of water. This homogenous mixture is used to prepare brick molds using a press. The green bricks are initially air-dried (1–2 days) then water cured for about 3 weeks
4	Fly ash cement bricks	Fly ash (50–60%), sand/stone dust (30–40%), and cement (8–10%) are mixed with required amount of water. The homogenous mixture is pressed and cured
5	High fly ash content bricks	Fly ash (90–94%) and caustic soda additives are mixed in required proportions and pressed to form green bricks. The product is cured in air for 4 weeks

quality of the bricks is specified by the Bureau of Indian Standards (BIS) vide code IS; 12894-2002. To encourage the use of bricks, Ministry of Environment and Forest MoEF, Government of India, has notified that all constructions within a radius of 100 km from power plant should be made using fly ash bricks with minimum 50% fly ash content in the brick (NTPC 2020). The quality of the ash bricks depends on the nature of ash, proportion of raw material, and their mixing, press, and curing. National Thermal Power Corporation, India (NTPC) recommends to use the ash from first or second field of the ESP for brick making. The quality is mainly dependent on the amount of cement used in the mixture (Table 3.6).

There are various technologies for manufacturing of bricks from fly ash. Steam curing of the bricks made from ash+sand+lime mixture requires high capital and operational cost. This technology is not popular. The other process requires low capital and recurring cost where curing of bricks (ash+lime+gypsum) is done using normal or steam process. Clay+fly ash molds followed by firing are also practiced; this process could be easily adopted by the existing clay fly ash brick manufacturing units.

Clay+fly ash brick has high market acceptability. Fly ash addition to clay decreases the weight of the clay brick by 15–20% and improves the thermal insulation. The other advantages are high durability, good finishing of different shapes, considerable

Table 3.6 Effect of the proportion of cement on the quality of fly ash bricks

S. No.	Fly ash	Sand	Cement	Lime	Gypsum	Strength	Water absorption (%)	Efflorescence	Reference
(A)	55	20	0	20	5	126 kg/cm^2	7.63	<10%	Kumar et al. (2014)
	52	20	3	20	5	141 kg/cm^2	6.06	<8%	
	50	20	5	20	5	152 kg/cm^2	5.41	<7%	
(B)	46	25	0	22	7	122.4 kg/cm^2	5.78	<11%	Kumar et al. (2015)
	44	25	2	22	7	142.5 kg/cm^2	4.62	<9%	
	42	25	4	22	7	148.3 kg/cm^2	4.13	<8%	
	40	25	6	22	7	154.5 kg/cm^2	3.57	<6%	

reduction of fuel used in firing, good bonding between the brick and mortar, termite resistance, and so on (Das and Segaran 2005).

Fernández-Pereira et al. (2011) prepared bricks from biomass (olive mill cake) gasification ash by extruding the ash (15–20%) and clay mixture through a conventional mold. The green bricks were initially air-dried and then heated in a furnace at high temperature. The high carbon content in the ash helps to prepare light porous bricks of high water-absorption capacity. Clay brick prepared by mixing 30% olive biomass fly ash showed good mechanical strength with porous structure (Pérez-Villarejo et al. 2020). Bottom ash of wood, pine-olive pruning, olive stone, and olive pomace were mixed with clay and fired at 1000 °C. The quality of the bricks is better with low bulk density, acceptable mechanical strength, and no hazard of leaching of potentially toxic metals from these bricks (Eliche-Quesada et al. 2017). Wood ash (35%) mixed with ceramic waste was also used for preparation of acceptable quality bricks (Kumar and Singh 2017). There are several studies that demonstrate the successful use of biomass-based ash for making bricks. However, the process requires high energy for firing the clay-ash molds at high temperature.

3.4.2 Block

Interlocking concrete paver blocks are used in many outdoor applications. They are pre-fabricated concrete of different shapes and dimensions used for pedestrian and lightweight vehicle movement. Cement and aggregates (fine and coarse) are mixed with required amount of water and poured in molds. The blocks are cured for about 28 days. Normally considerable portion of cement is replaced with fly ash in the paver block.

Karasawa et al. (2003) reported that 25% substitution of cement with fly ash produced paver block of acceptable quality; however, 40% replacement did not produce acceptable quality blocks. Fly ash was used up to 30% cement replacement in concrete (Kumar 2017). Pattnaik et al. (2018) observed higher compressive strength when 20% cement was replaced by equal proportion of fly ash and glass powder.

Partial replacement of cement with rice husk ash has been studied (Reddy et al. 2015). Different varieties of biomass ashes were also evaluated for their suitability for use in paver block preparation. Waste tea ash was used as a cement replacement in concrete paver block. About 20–40% replacement of cement with tea ash produced paver blocks suitable for use in pedestrian and garden areas (Djamaluddin et al. 2020). Milling and sieving of bagasse ash could increase the mechanical properties of the paver blocks prepared from bagasse ash. Addition of 10% ash along with 10% lime significantly increased the mechanical strength and durability of soil blocks (Alavéz-Ramírez et al. 2012).

There are several studies and in many places use of fly ash for manufacturing paver blocks of different grades is in practice. However, the quantity of fly ash used in paver block is significantly low when compared with the fly ash bricks. Increasing the quantity of ash in blocks decreases the quality and acceptability.

3.4.3 *Lightweight Aggregates*

Lightweight aggregates are prepared by sintering the fly ash. The sintered products are crushed into desirable sizes and used in concrete. These aggregates are used in high strength and high-performance concrete. Lightweight concrete is about 20% lighter and 20% stronger than the normal concrete. Further, it is possible to decrease the usage of cement by 20% (Kayali 2008). The advantages of lightweight concrete are depicted in Fig. 3.2.

Ultra-lightweight ceramsite was prepared from a mixture of fly ash:kaolin:diatomite (80:15:5) and additional 10% perlite. The mixture was pre-heated at 800 °C followed by sintering at 1220 °C for 10 min. The product has bulk density 340 kg/m^3; particle density 0.68 g/cm^3, and cylinder compressive strength 1.02 MPa (Liu et al. 2020). Cold-bonded fly ash aggregates were also prepared by Usanova (2020). Green mold prepared by mixing fly ash with a super-plasticizer was sintered in a furnace at 1250 °C. After cooling the mass was broken into angular aggregates. Concrete prepared from this light aggregate fulfils the requirements for use in structural applications (Kamal and Mishra 2019). Pelletized fly ash aggregates can also be used in structural concretes (Güneyisi et al. 2015;

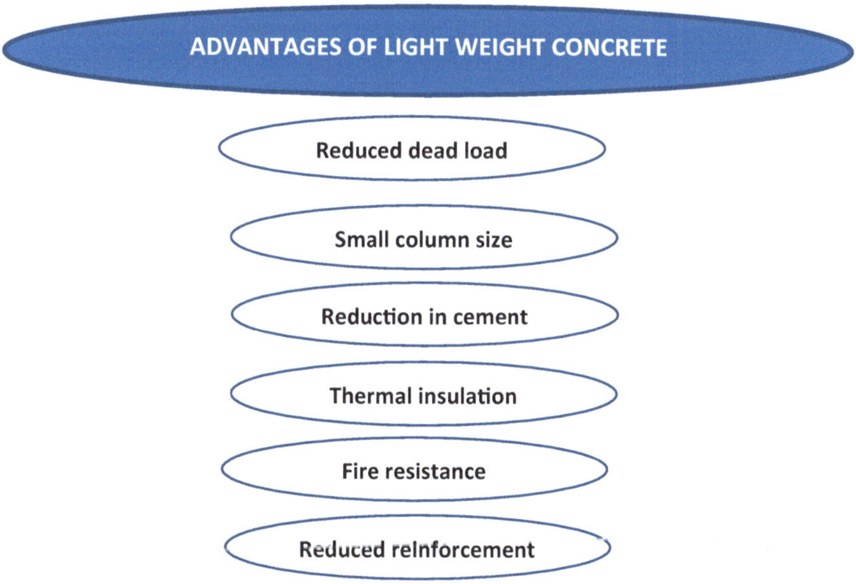

Fig. 3.2 Advantages of lightweight concrete

Kirubakaran and Joseravindraraj 2018). Huang et al. (2007) prepared lightweight aggregates from mixture of mining waste, incineration fly ash, and heavy metal sludge.

Biomass ash was also used for making lightweight aggregates. Lura et al. (2014) prepared lightweight aggregates by mixing bagasse ash with sodium bentonite, followed by granulation and sintering at 1100 °C. Moreno-Maroto et al. (2019) prepared lightweight aggregates from a mixture of biomass ash, beer bagasse, Zn-rich sludge, and clays. In another study, rice husk ash along with sodium hydroxide and boric acid was used (Chindaprasirt et al. 2009).

3.5 Ceramic Products

Ceramic tiles are typically produced from a mixture of clay (50%), feldspar (35%), and quartz (15%), where clay acts as a binder for other components. In green state, clay imparts the plasticity and helps in getting the desired shape. Low temperature melting feldspar generates the liquid that fills the pore space and produces a dense material. Quartz is generally stable at the firing temperatures and it helps to protect shrinkage and distortion (Luo et al. 2017). Depending on the nature of fly ash, it was used to replace a portion of clay or feldspar or quartz. Use of fly ash as a raw material for ceramics decreases the burden of the mining of the traditional raw materials. Zimmer and Bergmann (2007) prepared ceramic tiles using fly ash (up to 60% wt.). Addition of lime stone and feldspar increases the quality of fly ash tiles (Maitra 1999). Fly ash is used in a variety of ceramic products as indicated in Fig. 3.3.

Luo et al. (2017) prepared tiles by using raw fly ash and alkali-activated fly ash mixtures. The green mold obtained from the mixture of raw and activated ash was sintered at temperatures >1000 °C. Ceramic fibers and foams are prepared using fly

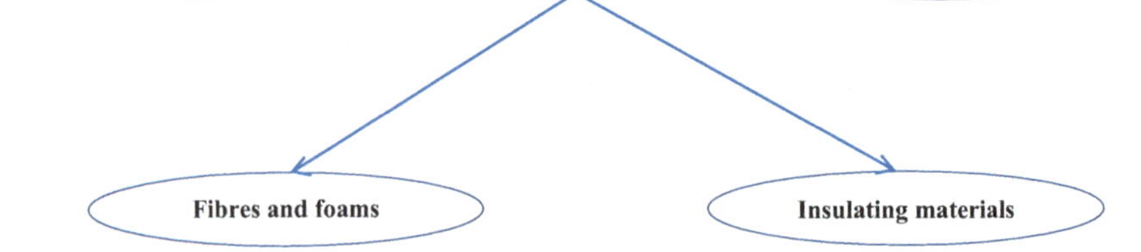

Fig. 3.3 Ceramic products prepared by using fly ash

ash, borax, and SiC. Fire clay bricks were prepared by mixing fly ash (10–50%) and clay followed by sintering at 1100–1400 °C (Sukkae et al. 2018). Use of fly ash in the preparation of refractories may decrease the linear shrinkage and apparent porosity of the refractory (Hassan and Aigbodion 2014) and possess good mechanical strength and low thermal conductivity (Otero et al. 2004). Thermal insulation materials were also prepared using different proportion of fly ash. A kind of sintered thermal insulation wall material was prepared by Yang et al. (2012) by using fly ash (65%). Lin and Chang (2013) prepared porous mullite ceramics with 20% fly ash. Several studies were conducted using different proportions of ash, the high sintering temperature is one of the key limitation for the economic viability of many of the insulating materials from fly ash (Zhang et al. 2014a). Zhang et al. (2020) prepared an additive-free glass ceramic from ash. Municipal solid waste incineration ash and coal ash mixtures were homogenized, melted, cooled, and powdered. This powdered parent glass material was molded and sintered at 950 °C.

Biomass ash is also used in the production of multi-purpose ceramic products. For getting quality ceramic products, the proportion of biomass ash used in the raw material mix is up to 20% (Pérez-Villarejo et al. 2012). Ash could be used for partial or full replacement of sand used in the raw material mix (Kizinievic and Kizinievic 2016). Roof tiles were prepared from sugarcane bagasse ash by mixing up to 60% ash in the mix (Souza et al. 2011). Washed olive pomace ash was used for preparation of fire clay bricks of acceptable quality (de la Casa and Castro 2014). Fire wood ash can be used up to 4.2 wt.% for making electrical siliceous porcelain material (Coutinho et al. 2019).

3.6 Road Construction

Both fly ash and bottom ash could be used in road construction. In road projects, fly ash is used as a substitute for cement in concrete/grout, fill material in embankment, aggregate for highway subgrades and road base, and in flowable fill; whereas, bottom ash is used as aggregate in concrete and in cold-mixed asphalt, and as a structural fill for embankments and cement stabilized bases (EPA 2005). The different uses of fly ash in road construction are depicted in Table 3.7. Construction of embankments using topsoil is expensive and causes environmental degradation. Projects like road construction and runways also use significant amount of fly ash. Fly ash has many favorable properties that help in its use for embankment like lightweight, high stability, moist, no lumps, and so on. Fly ash is easily available and the physical properties of the ash help in soil stabilization. Shear strength obtained using fly ash embankments is usually higher due to the self-hardening or pozzolanic properties of the ash (Ahmaruzzaman 2010). The pozzolanic properties of fly ash and the alkali provided by lime or cement are exploited for binding the soil. The initial strength is governed by the mechanical properties of the mixture; however, the ultimate strength and durability depend on the pozzolanic reactions.

Table 3.7 Application of fly ash in road construction (*Source* ACAA (2003))

	Application	Brief description
1	Cement concrete	Addition of fly ash to cement concrete improves the property of the concrete. Use of fly ash concrete has high ultimate strength and durability, better workability, low bleeding, low heat of hydration, decreased permeability, resistance to chemical attack, and decrease of cost. Use of high carbon containing fly ash may decrease durability
2	Stabilized base course	Road bases are generally prepared by mixing fly ash, lime, and aggregates. This is also used instead of cement. The main advantage of using ash is preparation of strong and durable base, low cost, energy efficiency, autogenous healing
3	Flowable fill	It is a liquid-like material comprising ash, cement, and water. No compaction or vibration is required to set. Sometimes other filler materials like sand, bottom ash, or quarry fines are included in the mixture. The flowable fill is a substitute for the traditional backfill materials like sand, gravel, and soil. The main advantages are no compaction is required; can be used in sites/structures that are not easily accessible to traditional fill techniques; increases soil bearing capacity, around/under structures inaccessible to conventional fill placement techniques; increases soil-bearing capacities; fast backfilling, avoids post-fill settlement issues, reduces heterogeneity in the density of the backfilled materials; more safety, less labor, and can be easily excavated later
4	Structural fills/embankments	Like flowable fill, fly ash is employed as a borrow material to construct fills and embankments. These structural fills are capable of supporting small road shoulders, highway buildings, and large embankments. This is cheap, avoids digging of surface soil, can be used in low soils with low bearing strengths, easy to handle with minimum construction equipment and with decreased construction time. Severe dust control and erosion prevention measures should be employed
5	Soil stabilization	Fly ash is used for stabilization of soils, particularly to improve soil density, strength, water content, and plasticity. The main advantage of using fly ash to improve soil stability includes the elimination of use of valuable borrow materials, increases construction speed by improving the quality of wet or unstable soil, reduces the requirement of pavement thickness, and decreases the requirement of natural aggregates in the pavement
6	Asphalt pavements	Fly ash is used as mineral filler in hot mix asphalt pavements. Fly ash improves the stiffness of the matrix, rutting resistance, and durability. Because of the presence of lime and the hydrophobic nature of the ash, the propensity for asphalt stripping is decreased
7	Grouts for pavement sub-sealing	Fly ash, water, and other materials mixtures are used as grouts for filling the voids under a pavement system without sub-sealing. It is also used to raise and support concrete pavements by injecting the grout. So, under mining is rectified using fly ash grouts without disturbing the surface pavement. This develops strength and can be done without much disturbance to the road traffic. However, some curing period is needed before using for heavy traffic. Confinement of the grout under the pavement has to be ensured

3.7 Mine Backfilling

The ash generated from coal combustion has found application to amend the environmental impacts of mining, through grouting or backfilling the gaps/voids created due to mining activities. During underground excavation exercises, backfilling of voids helps in maximizing ore extraction and stabilization of structures simultaneously. It is the spherical morphology that enables fly ash to be used as backfilling materials, as the sphericity allows it to flow and fill the voids. By regulating the fly ash-cement proportions in the mixture, grouts are prepared depending on the end applications. Owing to their binding properties and higher silica-alumina contents, fly ashes have been used to partly replace cementing materials during mine backfilling.

For amending open cast mining-induced-surface damages and underground mine backfilling, the application of coal ash produced from combustion has certain environment-friendly outcomes (Ram and Masto 2010), which include deterrence on acid-mine-drainage (AMD), agricultural surcharge, subsidence prevention at underground mine sites and improving land stability, and so on (Mishra and Das 2010; Sivakugan et al. 2006; Vories 2005). AMD typically is observed in areas where coal has been mined and the exposed sections or gangue piles contain pyritic substances. These pyritic components when come in contact with water and oxygen, react and induce AMD. Fly ash injection in underground mines has also found application for controlling mine fire episodes (Ahmaruzzaman 2010). The modus operandi for AMD prevention or land-subsidence prevention or fire control is similar.

Backfilling of mine-abyss is achieved either by using non-cemented or cemented backfilling mechanisms. Coal ash has found widespread curiosity and applications for its usage in cemented backfilling materials where binding agents (3–6%) such as fly ash or Portland cement are used to enhance the stability (Sivakugan et al. 2006). The most encouraging part about the prospect of using coal ash in backfilling is that it creates the possibility of bulk utilization of ash generated from combustion. Dry injections (at low pressures of ~200 kPa) are achieved using ~6 in. diameter boreholes drilled at mine-voids and injecting the coal ash which takes up moisture from adjoining strata and consequently hardens. Wet injections (at higher pressures of ~700 kPa) employ distribution of fly ash and water slurries (Ahmaruzzaman 2010), using ~10 in. diameter boreholes through which the slurry is injected. As a prerequisite, the water-ash slurry should have good flowage properties, readily reach the voids, and harden in due course of time to prevent land subsidence. Normally, the above-mentioned two components are present in 1:1 ratio, or 0.8:1 ratio (Twardowska 1999). Wang et al. (2009) used coal gangue in addition to fly ash to partly substitute cement during backfilling of a coal mine. It was observed that the addition of fly ash along with some other composites facilitated the flow and movement of the slurry through the pipelines. Similarly, in Australia too, fly ash has been used to partly replace cement during backfilling, and it has been reported that it facilitated uninterrupted longwall mining (Grice et al. 1999).

The overall possibility of using fly ash in mining sector is depicted in Fig. 3.4. Paste backfilling is a relatively new technology to fill the mining voids. The paste

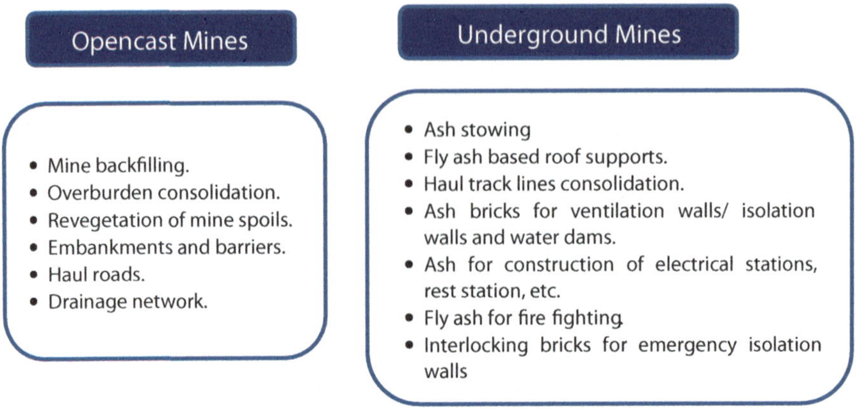

Fig. 3.4 Use of fly ash in mining sector (Adapted from Rao and Kumar (2019))

contain 70–85% solids (Fall and Pokharel 2010). Cement is normally used as a binder in the paste. Barricades are constructed at the base of the slopes to contain the backfill within the zone (Helinski et al. 2011). Paste fill has excellent ground control, reduces mining cycle time, avoids segregation problem, uses less binder and water, subsidence and fire control, improves ventilation, and decreases ore dilution (Célestin and Fall 2009; Li and Aubertin 2008). Fly ash and mine water, as the major components of the paste fill, were demonstrated in Zhaoguan coal mine. The results showed that the filling rate was 43.46% and the surface subsidence coefficient of the grouting process was 0.475 (Jiang et al. 2017). Rani and Jain (2017) reported that 50% slurry consisting of fly ash and bottom ash mixture (7:3, 6:4, and 5:5) has the favorable properties for use in mine filling.

Fly ash paste filling has three steps as detailed in Table 3.8.

Table 3.8 Steps involved in fly ash paste filling (*Source* Prashant et al. (2005))

	Steps	Brief detail
1	De-watering	Coal ash is generally discharged as dilute slurry in ash pond. So thickening or de-watering is required. Conventional gravity thickener is often used to generate slurry of 65–75% solids
2	Mixing	A screw mixer is used to mix the components of the paste fill. Filter cake, binder, and water are accurately weighted and mixed either through a batch or continuous mixing system
3	Pumping, transportation, and deposition	High-volume positive displacement pumps are used for conveying the paste fill mixture into the underground void or surface impoundments. This can also be conveyed through gravity or gravity + pump combination

The advantages and disadvantages of paste filling were elaborated by (Prashant et al. 2005).
Advantages:

1. Low permeability, hence cement washout and cement requirement are low.
2. Cost of mine de-watering is decreased; pozzolanic properties of ash help in solidification of the paste.
3. Higher strength with less cement.
4. All kinds of ash (fly/bottom/pond) could be used; but for hydraulic fill only coarse bottom ash is used.
5. Dense paste, so more confinement strength.
6. Enhanced safety.
7. Decreased dilution in weak ore body.
8. Low water content, thereby low investment on the underground confining drainage barricades.

Disadvantages:

1. High capital cost.
2. Paste pumping depends on the water content and grain size distribution.
3. Increased pressure pipelines and greater level of engineering of piping network in the mines is required.

3.8 Extraction of Value-Added Materials

3.8.1 Cenosphere

Cenospheres are hollow spheres present in fly ash and their chemical composition is similar to the fly ash. They have spherical shape with low bulk density, high thermal resistance, high workability, and high strength (Ranjbar and Kuenzel 2017). They have many industrial applications (Fig. 3.5). As the density of cenospheres is less than water, they are generally extracted from the surface waters of the fly ash lagoons. However, this process is time-consuming and space-consuming and continuous extraction is not possible. Gurupira et al. (2001) suggested an alternate turbo-electric separation system, where cenosphere is defined as particles having specific gravity <2. However, for many applications, cenosphere with specific gravity <1.0 is essential. The performance of the lightweight products depends on the cenosphere diameter, particle size distribution, wall thickness, and shape. A float sink method is also employed for cenosphere separation. Ash is dispersed in water, and after stirring it will be allowed to settle for 24 h. The floats are separated and heated at temperature around 800 °C to remove organic materials. The yield of cenosphere may vary from 0.1 to 5.0% (Joseph et al. 2013). Both dry and wet methods are available for separation of cenosphere from ash. As the yield of cenosphere is low,

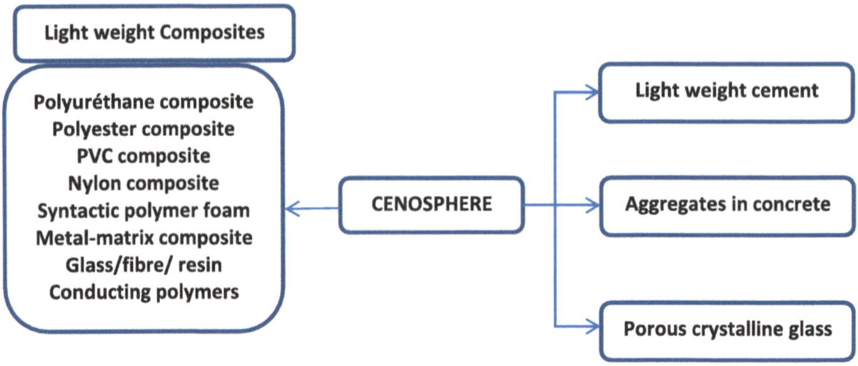

Fig. 3.5 Applications of cenosphere

attempts were also made for synthesis of cenosphere by thermal flame projection process (Saucedo-Salazar et al. 2014).

The mechanism of cenosphere formation is detailed by Ranjbar and Kuenzel (2017) and Danish and Mosaberpanah (2020). During combustion of pulverized coal in the boiler, the silicate and other minerals present in the coal get decomposed resulting in the formation of more glassy amorphous phases and few crystalline phases. Due to further heating, there exist molten and non-molten phases, where the non-molten quartz is captured by the molten amorphous phases which partially dissolve quartz leading to high SiO_2/Al_2O_3 ratio. This alters the surface tension and viscosity of the molten liquid. Merging of individual micro droplets into larger droplets and gases formed due to the decomposition of organic and inorganic matter inflates these larger droplets into hollow spherical cenospheres. The properties of the cenosphere depend on the molten fluid property and the gas pressure. The balance between the gas pressure and the surface tension pressure of the sphere determines the stability of the cenosphere. Rapid cooling decreases the gas pressure inside the sphere, if the surface tension yields to gas pressure; the hollow sphere blows up into many dense or small hollow spheres. High viscous melts (Al-SiO_2, non-magnetic) form larger spheres with thin wall; low viscous melts (FeO, magnetic) cannot withstand gas pressure and form small porous shell structures. Ngu et al. (2007) found more SiO_2/Al_2O_3 in smaller cenosphere than large cenosphere; smaller cenospheres have single-ring structure, whereas network structure in large cenospheres.

Some biomass ashes having high amount of silica (like rice straw ash) may be exploited for cenosphere extraction by following sink-float separation (Vassilev et al. 2013b).

3.8.2 Unburnt Carbon

Depending on the combustion efficiency, the unburnt carbon content in the fly ash may vary from 2 to 15%. Presence of unburnt carbon is a limitation for use of fly ash especially in cement and concrete. Unburnt carbon in ash are formed in three stages; initially the moisture content in the coal is removed, followed by combustion of organic matter leading to increase in ash content and eventually the oxygen content decreases which retards subsequent burning of unburnt carbon particle in ash. FTIR studies showed the presence of oxygen-based functional groups on the unburnt carbon surface which decreases the floatability of carbon while separation. Unburnt carbon particles are porous with high surface area. Different methods of separation of unburnt carbon from ash are presented in Table 3.9.

Commercial-grade carbon materials are being extracted in many countries with carbon content up to 70% (Baltrus et al. 2001). Unburnt carbon separated from lignite ash was used for co-combustion with lignite (Rybak et al. 2017).

Unburnt carbon in biomass ash can be up to 70%, normally 1–20%. There are several applications for the carbon extracted from coal or biomass ash (Fig. 3.6). The carbon extracted from biomass ash has properties comparable with the commercial activated carbon in terms of surface area, pore volume, surface functional groups, and adsorption capacity (Vassilev et al. 2013b).

Table 3.9 Methods for extraction of unburnt carbon from fly ash

	Methods	Brief details
1	Sieving	It is easier when the carbons are present in the large-sized ash particles. The efficiency of this method is low, so it can be used as a pre-treatment process before employing other separation techniques
2	Gravity separation	Density difference between mineral and carbon particles is exploited in this technique
3	Electrostatic separation	It is based on bipolar charging of dry particles either by particle–particle contact or by particle–wall collision under turbulent conditions. Then it is possible to separate the positively charged carbon particles from the negatively charged ash particles in an electric field
4	Froth floatation process	Highly versatile method. It is based on the adherence of carbon particles on the air bubbles and they are positively buoyant and reach the top of the slurry. Hydrophobic nature of carbon aids in floatation
5	Oil agglomeration	Efficient for separation of micro fine particles. Wet carbon particles form agglomerates while the mineral particles remain in the suspension

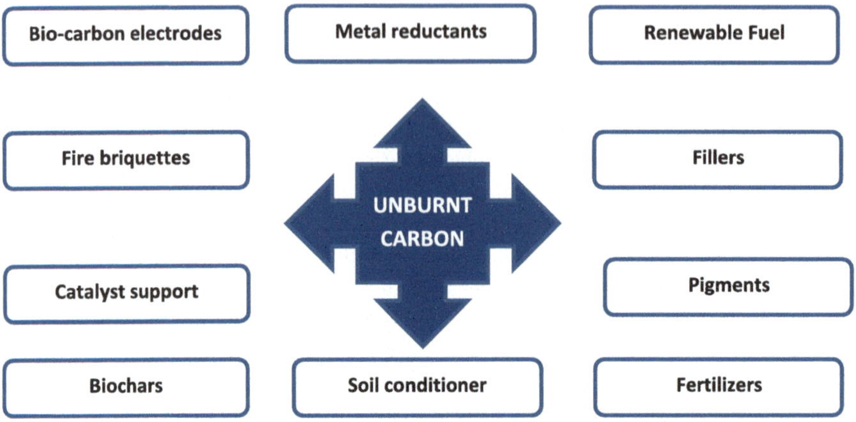

Fig. 3.6 Applications of unburnt carbon extracted from ash

3.8.3 Extraction of Rare Earth Elements

Some fly ash contains economically extractable amounts of rare earth elements. However, the extraction technologies are complex that limit this application. Extraction of rare earth elements from fly ash has several advantages than its extraction from ore. Fly ash is readily available and no expensive excavation is needed, already in fine particle size, no generation of unwanted rocks and mine spoil. Presence of significant amount of rare earth elements in coal ash has been reported from UK (Blissett et al. 2014), China (Dai and Finkelman 2018), Indonesia (Rosita et al. 2020), and many countries. Taggart et al. (2016) reported significant potential for commercial extraction of rare earth elements from US coal ashes, especially from the Powder River Basin ashes. However, success depends on the development of suitable extraction technology. The different techniques of separation of elements from ash are presented in Table 3.10.

Understanding of the chemical association of the rare earth elements helps in designing suitable extraction method. In coal ash, these elements could be associated with their respective oxides, phosphates, apatite, zircon, and in glass phase as discrete particles or uniformly distributed in the glass phase (Liu et al. 2019). Content of rare element is more in fly ash than bottom ash. These elements are not enriched in any size fractions of fly ash and are uniformly distributed in all size fractions (Ma et al. 2019b). Sequentially, extraction showed the following order of association of rare earth elements: silicate-aluminate > organic or sulfide > acid soluble > metal oxides > ion-exchangeable form (Pan et al. 2018).

Lithium is enriched with some coal and fly ash. CO_2 is pumped into the ash slurry to convert lithium carbonate into more soluble lithium bicarbonate and filtered. The filtrate is decarbonated to precipitate pure lithium carbonate (Grantham and Yosim 1974). During coal combustion gallium and germanium get condensed on the surface of fly ash particles and therefore enriched in ash than the coal. Several techniques

Table 3.10 Different techniques for extraction of elements (Adopted from Gollakota et al. (2019))

	Extraction technique	Brief details
1	Sublimation	Process involves direct transition from solid to gas phase bypassing the liquid phase. A temperature gradient and a reduced partial pressure are employed generally for purification of high-grade materials
2	Magnetic separation	Widely applied for extraction from and separation of Fe containing materials. An external magnetic field is applied for separation. Both wet and dry magnetic techniques are employed
3	Chlorination	The non-magnetic fraction of the ash is chlorinated in the presence of carbon to get pure metals
4	Fractional crystallization and precipitation	Fractional crystallization involves precipitation of salts from solution by varying the temperature of the saturated solution. Regarding fractional precipitation, rare earth elements in solution are partially separated in the form of a less soluble compound by using chemical additives
5	Ion exchange	This technique uses ion exchange resins to prepare high pure rare earth elements from impure mixtures. Elective adsorption is achieved by changing the pH and concentration of the solution
6	Extraction and leaching	Both acidic and alkaline extracting solutions are used for dissolution of rare earth elements from coal ash. The element from the resulting solution is separated either by distillation or evaporation or crystallization
7	Electrochemical process	The targeted metal ions in the solution are deposited on the electrode surface under the influence of electric current
8	Sintering	Thermal treatment of the ash below its melting point which results in bonding of particles with increased strength. Lime sintering is heating of ash along with lime at high temperature. The resultant product is dissolved in alkali to separate Al from the insoluble calcium silicates. In calsintering ash-lime-gypsum mixture is sintered followed by acid leaching to separate valuable elements

available for extraction of these elements are reviewed by Sahoo et al. (2016). Coal ash was digested by hydrothermal process and an XAD-7 resin was used for separation of rare earth elements from the digested solution (Mondal et al. 2019). Recently, bacterial strains were used for bio-sorption of rare earth elements from coal ash (Park et al. 2020). In biomass ashes, the concentration of rare earth elements depends on the nature of the plant species. The concentration of rare earth elements in biomass ash is much lower than coal ash. However, presence of many water-soluble components in biomass ash may help in co-extraction of these elements with other components (Vassilev and Vassileva 2020).

3.8.4 *Aerogels*

Aerogels is another recent application of fly ash. Presence of Si and Al with excellent insulation properties favors the use of fly ash as a starting material for the synthesis of aerogels (Gollakota et al. 2019). Aerogels has special properties like high porosity, low density, high surface area, low dielectric constant, low thermal conductivity, low optical index of refraction, and high optical transmission, elasticity, durability, and flexibility. Several studies support the use of fly ash for the synthesis of aerogels for different applications: adsorption of drugs (Smirnova et al. 2003); mercury cleanup (Graham and Thomas 1999); Cd adsorption (Givianrad et al. 2013); catalyst (Li et al. 2016); insulation material (Lu et al. 2018); and fire safety (Kang et al. 2017). The different applications of aerogels are presented in Fig. 3.7.

For preparation of good quality aerogels, fly ash has to be pre-treated to remove the impurities present in the ash. Initially the ash is calcined for about 3 h at 750 °C followed by acid treatment to ease the separation of Si from ash. Hydrolysis reactions lead to the formation of silanol groups and condensation of these silanol forms colloidal gel structures through siloxane linkages, and thereby forming a continuous interconnected network of Si-O–Al–O structures. Aging of these gels make them rigid and cohesive. Air drying of these gels leads to contraction and fracture due to the capillary forces (Gollakota et al. 2019). Wu et al. (2018) prepared aerogels by using fly ash and trona ore; the final product has high thermal stability and large specific surface area. Sugarcane bagasse ash was also used for preparation of aerogels (Nazriati et al. 2014). Other biomass ash, like ground nut shell ash (Guzel Kaya and Deveci 2020) and bean pod ash (Kaya et al. 2019), has been used for preparation of aerogels.

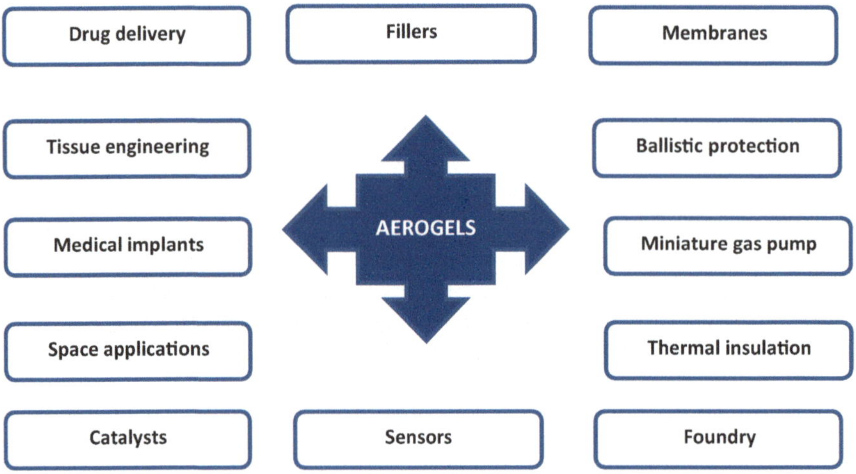

Fig. 3.7 Different uses of silica aerogels (Modified from Gollakota et al. (2019))

3.8.5 Zeolite

Extensive studies were carried out for synthesis of zeolite from coal and biomass ash. Presence of reactive aluminosilicates in ash and high specific surface area make it an amicable raw material for zeolite synthesis. It is prepared by hydrothermal crystallization of ash in the presence of strong alkali. Presence of quartz and mullite in significant amount may be one of the limitations for fly ash use for zeolite synthesis. According to Yao et al. (2015), there are three steps in the preparation of zeolite synthesis: (1) dissolution, (2) condensation, and (3) crystallization. Use of $Li(OH)_2 \cdot H_2O$ produces highly crystalline zeolite (Yao et al. 2009). Hollman et al. (1999) proposed a two-step process where initially Si is extracted to adjust the Si/Al ratio; by this method improved crystalline zeolite was obtained. Some of the approaches for zeolite preparation are presented in Table 3.11.

Zeolites are mostly used as an adsorbent, for treating waste water, cleaning gas, and as a substitute for phosphate in detergent. Large pore size zeolites with high CEC are used as molecular sieves and for ion exchange (Ahmaruzzaman 2010). Long reaction time and consumption of water are the main disadvantages of using fly ash for zeolite synthesis. High temperature process with water removal during the process was introduced by Moriyama et al. (2005).

Regarding biomass ash, mostly rice husk ash was used for preparation of zeolite for various purposes: catalyst (Chareonpanich et al. 2004; Krisnandi et al. 2017; Mohamed et al. 2015; Naskar et al. 2012). There are several studies on the preparation

Table 3.11 Different processes used for synthesis of zeolite from ash

	Process	Details
1	Hydrothermal process	The process involves hydrothermal treatment of ash with alkali (2–13 M) at elevated temperature (80–100 °C) and varying time (8–48 h); followed by centrifugation, washing, and drying of the product
2	Alkaline fusion-assisted hydrothermal process	Similar to hydrothermal process with an additional pre-treatment step of fusion with alkali at temperature around 500 °C
3	Multi-step treatment	Hydrothermal process where additional steps were included for pre-addition of salts or leaching of fly ash to separate the soluble silica, adjusting the Si/Al ratios, etc.
4	Microwave heating	Hydrothermal process where the ash-alkali mixture is heated in a microwave. The main advantage is reduction in the activation time from several hours to 30 min
5	Sonication approach	Ultrasonic treatment accelerates solubility of amorphous Si and Al, enhances the condensation process bond strength, and finally the crystallinity of the product

of zeolites from rice husk (Dalai et al. 2005; Dey et al. 2013; Ghasemi et al. 2011; Mohamed et al. 2015; Ng et al. 2015; Wang et al. 2018b, c), wheat straw (Ali et al. 2018), and bagasse ash (Assawasangrat et al. 2016; Moisés et al. 2013; Purnomo et al. 2012; Shah et al. 2011).

3.9 Niche Areas of Ash Utilization

3.9.1 Catalyst

Use of fly ash in catalytic applications, especially in heterogeneous catalysis, has been studied in detail. Fly ash is used as a component of active catalyst as well as catalyst support material in heterogeneous catalyst due to the presence of Al_2O_3 and SiO_2 and favorable thermal stability for its use. Some of the studies on fly ash use as a catalyst are detailed in Table 3.12.

Most of the fly ash-based catalysts are microporous. Fly ash-based catalysts are widely used in degrading pollutants present in gas and water, chemical conversion reactions, and manufacturing of chemical substances. Use of fly ash increased the conversion/reaction efficiency; these catalysts can be regenerated and used for multiple cycles without change in the structure (Asl et al. 2018).

3.9.2 Adsorption of Toxic Elements and Organic Compounds

Fly ash can be used as a low-cost adsorbent for removal of pollutants from gas and water (Ahmaruzzaman 2009). Fly ash properties like morphology, surface area, porosity, and Si–Al content make it a suitable material for adsorption and waste water remediation (Mushtaq et al. 2019). Modified Class 'F' fly ash was used for removing the Pb^{2+}, Cd^{2+}, and Zn^{2+} cations from mixed solutions (Visa et al. 2012). Toxic phenolic compounds could be removed by fly ash (Akgerman and Zardkoohi 1996). The adsorption efficiency of different organic compounds by fly ash is shown in Fig. 3.8.

Fly ash is also used for capturing Hg (Ochedi et al. 2020; Wang et al. 2016; Yang et al. 2017) and As (Zhang et al. 2019) from flue gas and water. Generally, activated carbon is used for removal of Hg and As, but use of activated carbon is extremely costlier. Presence of unburnt carbon and other oxides, especially Fe oxides in ash, helps in oxidation and absorption of Hg0 (Yang et al. 2017). Addition of Co, Mn, and Fe oxides to fly ash enhances the removal of Hg0 from flue gas (Xing et al. 2012; Xu et al. 2014). Fly ash modification increased Hg0 removal in the following order: HBr > $CaCl_2$ > $CaBr_2$ (Zhang et al. 2014b). Modified fly ash with metal oxides and halides enhanced the removal of Hg0 from flue gas (Yang et al. 2016). NH_4Br modified fly ashes (Gu et al. 2015) and zeolite synthesized from fly ash also increased

Table 3.12 Use of fly ash in catalytic reactions

	Fly ash use in catalyst	Application	Reference
1	Support material for nickel	CO_2 methane reforming	Wang and Lu (2007)
2	Support for Fe, Cu, Ni, and V catalysts	Selective reduction of NO by ammonia	Xuan et al. (2003)
3	TiO_2 coated coal fly ash	Removal of NO from gas stream	Yu (2004)
4	Fe^{3+} loaded fly ash	Activation of H_2O_2 for use in advanced oxidation process	Flores et al. (2008)
5	Nano-crystalline catalyst prepared by loading sulphated zirconia on fly ash by sol–gel technique	Liquid phase benzylation of benzene and toluene with benzyl chloride	Khatri et al. (2010)
6	Activated fly ash as a support for Ru catalyst	Decomposition of ammonia	Li et al. (2008)
7	NaOH treated and thermally activated fly ash catalyst	Condensation reaction of benzylaldehyde and cyclohexanone to dibenzylidene-cyclohexanone	Jain et al. (2011)
8	Fly ash as one of the active components of catalyst	Oxy-chlorination and the deep oxidation reaction of phenol and mono-chlorophenols	Born et al. (1993)
9	Fly ash-saw dust	Degradation of 2-chlorophenol	Ma et al. (2014)
10	Co-fly ash	Phenol degradation	Saputra et al. (2012)
11	CaO-fly ash	Trans-esterification of soybean oil	Chakraborty et al. (2010)
12	Acid modified fly ash	Degradation of polyacrylamide	Wang et al. (2020)
13	Ni supported on alkali treated fly ash	Biogas dry reforming reaction	Gao et al. (2020)
14	Mesoporous silica from fly ash as a support material for Co–Fe bimetallic catalyst	Fischer–Tropsch synthesis	Gupta et al. (2020)
15	Fly ash-based zeolites	Gas-phase alkylation of phenol	Gjyli et al. (2019)

Hg removal from flue gas (Ma et al. 2019a; Wang et al. 2015). Magnetic modification also increased the Hg removal by fly ash (He et al. 2015). Non-thermal plasma was also used to modify fly ash to enhance Hg adsorption (Shi et al. 2019; Zeng et al. 2017).

Like Hg, fly ash has been used for removal of As from flue gas (Fan et al. 2019; Zhang et al. 2016b). Fe-enriched fly ash (50% Fe_2O_3) greatly enhanced As removal than the raw ash. Zhang et al. (2016a) separated ferrospheres from fly ash and used for As removal from flue gas. Raw or modified fly ash has wide applications in water

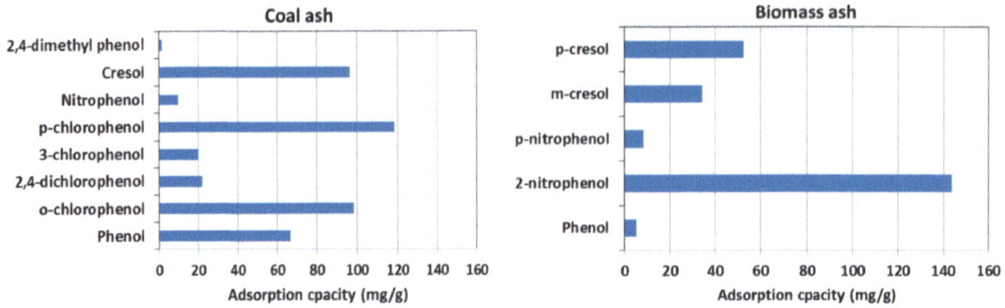

Fig. 3.8 Adsorption capacity of coal and biomass ash (Modified from Ahmaruzzaman (2009))

treatment for removal of several toxic elements and organic pollutants (Mushtaq et al. 2019).

3.9.3 Fertilizer from Ash

Apart from Si and Al, fly ash contains many essential elements required for plant growth. Processing of the Al-silicate phase of the ash helps in controlled release of the added fertilizer. Magnetized fly ash-based compound fertilizer was prepared by Dindi et al. (2019). This fertilizer increased crop yield; N and K use efficiency. Drum coating technology was used to prepare large granular urea where urea is coated by double layer; inner coating with attapulgite-fly ash and outer with Eudragite. The nutrient release pattern of the fertilizer was excellent; 65.74% release in 28 days (Fu et al. 2019). Alkali-activated fly ash has a matching pore size to hold triple superphosphate fertilizer. This fly ash-phosphate fertilizer is a slow release P fertilizer. Fly ash organic material mixture-based composted fertilizers were also prepared by many researchers (Mupambwa et al. 2015). Alkali-activated fly ash was mixed with NPK fertilizer and granulated to prepare slow release fertilizer. Silica fertilizer was prepared by activation of the ash residue obtained after acid leaching to remove Al (Wang et al. 2019). Fly ash-based slow release-coated fertilizer was prepared by Qi et al. (2019). Charcoal, fly ash, and nitrification inhibitor containing slow release fertilizer was prepared by Gao and Hu (2018). Potassium zeolite fertilizers were prepared from coal ash (Elliot and Zhang 2005; Flores et al. 2017).

Nitride-based slow release fertilizers were prepared by carbo-thermal reduction of fly ash. The N content in the product was up to 28%. The total energy required for the nitride process is comparable with the conventional Haber–Bosch process (Qiu and Hlavacek 2010). Amonette and Matyas (2018) prepared a controlled-release solid nitride fertilizer using fly ash. The product consists of mixture of solid nitride with precipitation-inhibiting agent and moisture-retention agent. This fertilizer releases N uniformly over the crop growth period and can increase the N use efficiency. Fly ash pelletized with KOH or KCl was sintered at 400 and 600 °C to prepare a slow release potassium fertilizer (Mario 2006). In another study, fly ash, potassium carbonate,

Table 3.13 Process for CO_2 capture by using fly ash

	Process	Details
1	Wet process	The alkali components of the ash and the added alkalis react with CO_2 to form their respective carbonates
2	Dry process	Fly ash is used as a support material for the sorbent like amine; as well as raw material for preparation of zeolites-based sorbents. These sorbents are mostly regenerated

KOH, coal, and other additives are granulated and calcined to prepare potassium silicate fertilizer (Segawa and Akizuki 1981; Shin et al. 1994). Granulated biomass ash (Clarholm 1994) and spent wash ash (Suhas et al. 2019) were also advocated as valuable fertilizers.

3.9.4 Carbon Sequestration

Many of the carbon capture technologies are expensive and have energy-intensive regeneration process, especially for adsorption. However, fly ash can provide a cheaper solution for carbon capture. Fly ash mixed with CaO, MgO, zeolites, Al_2O_3 was used for CO_2 capture. Both wet and dry processes are employed for CO_2 capture (Table 3.13).

Presence of significant amount of alkalies in ash makes it as a suitable material for CO_2 capture. Class C fly ash is more suitable for CO_2 capture due to the high content of CaO. Different chemicals were added to fly ash to increase CO_2 capture. The other advantage of this process is the abatement of toxicity by co-precipitation of potentially toxic elements along with the carbonates. Several biomass ashes have showed excellent CO_2 capture due to the presence of more amounts of alkalies in biomass ash than coal ash. Apart from raw fly ash, modified fly ash was also used for CO_2 capture and storage (Dindi et al. 2019).

References

ACAA (2003) Fly ash facts for highway engineers. American Coal Ash Association. US Department of Transportation, Federal Highway Administration

Adamu M, Mohammed BS, Shafiq N, Liew MS (2018) Durability performance of high volume fly ash roller compacted concrete pavement containing crumb rubber and nano silica. Int J Pavement Eng, 1–8

Adjei-Nsiah S (2012) Response of Maize (Zea mays L.) to different rates of palm bunch ash application in the semi-deciduous forest agro-ecological zone of Ghana. Appl Environ Soil Sci, 870948

Adriano D, Woodford T, Ciravolo T (1978) Growth and elemental composition of corn and bean seedlings as influenced by soil application of coal ash 1. J Environ Qual 7:416–421

Adriano DC, Page AL, Elseewi AA, Chang A, Straughan IA (1980) Utilization and disposal of fly ash and other coal residues in terrestrial ecosystem: a review. J Environ Qual 9:333–344
Adriano DC, Weber JT (2001) Influence of fly ash on soil physical properties and turfgrass establishment. J Environ Qual 30:596–601
Ahmaruzzaman M (2009) Role of fly ash in the removal of organic pollutants from wastewater. Energy Fuels 23:1494–1511
Ahmaruzzaman M (2010) A review on the utilization of fly ash. Prog Energy Combust Sci 36:327–363
Akgerman A, Zardkoohi M (1996) Adsorption of phenolic compounds on fly ash. J Chem Eng Data 41:185–187
Alavéz-Ramírez R, Montes-García P, Martínez-Reyes J, Altamirano-Juárez DC, Gochi-Ponce Y (2012) The use of sugarcane bagasse ash and lime to improve the durability and mechanical properties of compacted soil blocks. Constr Build Mater 34:296–305
Ali MM, Ahmed M, Hameed B (2018) NaY zeolite from wheat (Triticum aestivum L.) straw ash used for the adsorption of tetracycline. J Clean Prod 172:602–608
Amin N-U (2011) Use of bagasse ash in concrete and its impact on the strength and chloride resistivity. J Mater Civ Eng 23(5):717–720
Amonette JE, Matyas J (2018) Nitride-based controlled-release fertilizers and process for making same. Google Patents
Asl SMH, Ghadi A, Baei MS, Javadian H, Maghsudi M, Kazemian H (2018) Porous catalysts fabricated from coal fly ash as cost-effective alternatives for industrial applications: a review. Fuel 217:320–342
Assawasangrat P, Neramittagapong S, Pranee W, Praserthdam P (2016) Methanol conversion to dimethyl ether over beta zeolites derived from bagasse fly ash. Energy Sour Part A Recover Utili Environ Eff 38:3081–3088
Augusto L, Bakker MR, Meredieu C (2008) Wood ash applications to temperate forest ecosystems—potential benefits and drawbacks. Plant Soil 306:181–198
Bååth E, Arnebrant K (1994) Growth rate and response of bacterial communities to pH in limed and ash treated forest soils. Soil Biol Biochem 26:995–1001
Baltrus JP, Wells AW, Fauth DJ, Diehl JR, White CM (2001) Characterization of carbon concentrates from coal-combustion fly ash. Energy Fuels 15:455–462
Basu M, Pande M, Bhadoria PBS, Mahapatra SC (2009) Potential fly-ash utilization in agriculture: a global review. Prog Nat Sci 19:1173–1186
Behak L (2016) Soil stabilization with rice husk ash. Rice Technol Prod
Berra M, Mangialardi T, Paolini AE (2015) Reuse of woody biomass fly ash in cement-based materials. Constr Build Mater 76:286–296
Bhattacharya SS, Iftikar W, Sahariah B, Chattopadhyay GN (2012) Vermicomposting converts fly ash to enrich soil fertility and sustain crop growth in red and lateritic soils. Resour Conserv Recycl 65:100–106
Bi Y, Li X, Christie P, Hu Z, Wong M (2003) Growth and nutrient uptake of arbuscular mycorrhizal maize in different depths of soil overlying coal fly ash. Chemosphere 50:863–869
Biedermann F, Obernberger I (2005) Ash-related problems during biomass combustion and possibilities for a sustainable ash utilization. Austrian Bioenergy Centre GmbH, Bios Bioenergiesysteme GmbH.
Blissett RS, Smalley N, Rowson NA (2014) An investigation into six coal fly ashes from the United Kingdom and Poland to evaluate rare earth element content. Fuel 119:236–239
Born JG, Mulder P, Louw R (1993) Fly ash mediated reactions of phenol and monochlorophenols: oxychlorination, deep oxidation, and condensation. Environ Sci Technol 27:1849–1863
Bougnom BP, Knapp BA, Etoa F-X, Insam H (2011) Possible use of wood ash and compost for improving acid tropical soils, recycling of biomass ashes. Springer, pp 87–105
Bouzoubaâ N, Lachemi M (2001) Self-compacting concrete incorporating high volumes of class F fly ash: preliminary results. Cem Concr Res 31:413–420

Bundt M, Krauss M, Blaser P, Wilcke W (2001) forest fertilization with wood ash: effect on the distribution and storage of polycyclic aromatic hydrocarbons (PAHs) and polychlorinated biphenyls (PCBs). J Environ Qual 30:1296–1304

Campbell D, Fox W, Aitken R, Bell L (1983) Physical characteristics of sands amended with fly ash. Soil Res 21:147–154

Carlson CL, Adriano DC (1993) Environmental impacts of coal combustion residues. J Environ Qual 22:227–247

Célestin JCH, Fall M (2009) Thermal conductivity of cemented paste backfill material and factors affecting it. Int J Min Reclam Environ 23:274–290

CFRI (2001) Utilisation of fly ash in agriculture at Neyveli Lignite Corporation, Neyveli. CFRI, Dhanbad, Jharkhand. CFRI, Central Fuel Research Institute, 2001 Report No. TR/CFRI/3.01/2000–2001

Chakraborty R, Bepari S, Banerjee A (2010) Transesterification of soybean oil catalyzed by fly ash and egg shell derived solid catalysts. Chem Eng J 165:798–805

Chang A, Lund LJ, Page AL, Warneke J (1977) Physical properties of fly ash-amended soils. J Environ Qual 6:267–270

Chareonpanich M, Namto T, Kongkachuichay P, Limtrakul J (2004) Synthesis of ZSM-5 zeolite from lignite fly ash and rice husk ash. Fuel Process Technol 85:1623–1634

Chindaprasirt P, Jaturapitakkul C, Rattanasak U (2009) Influence of fineness of rice husk ash and additives on the properties of lightweight aggregate. Fuel 88:158–162

Christopher RS, Kimberly EK (2015) Use of biomass and co-fired fly ash in concrete. ACI Mater J 112:209–218

Clarholm M (1994) Granulated wood ash and a 'N-free' fertilizer to a forest soil—effects on P availability. For Ecol Manage 66:127–136

Cordeiro GC, Toledo Filho RD, Tavares LM, Fairbairn EMR (2008) Pozzolanic activity and filler effect of sugar cane bagasse ash in Portland cement and lime mortars. Cem Concr Compos 30:410–418

Coutinho NC, Loiola RL, Paes HR Junior, Holanda JNF (2019) Use of firewood ash waste in electrical siliceous porcelain. Mater Res, 22

Cruz NC, Rodrigues SM, Carvalho L, Duarte AC, Pereira E, Römkens PFAM, Tarelho LAC (2017) Ashes from fluidized bed combustion of residual forest biomass: recycling to soil as a viable management option. Environ Sci Pollut Res 24:14770–14781

Dai S, Finkelman RB (2018) Coal as a promising source of critical elements: progress and future prospects. Int J Coal Geol 186:155–164

Dalai AK, Pradhan NC, Rao MS, Gokhale K (2005) Synthesis and characterization of NaX and Cu-exchanged NaX zeolites from silica obtained from rice husk ash

Danish A, Mosaberpanah MA (2020) Formation mechanism and applications of cenospheres: a review. J Mater Sci 55:4539–4557

Das A, Segaran R (2005) Review of fly ash brick making technologies. J CAII (Coak Ash Institute of India) 7:369–404

de la Casa JA, Castro E (2014) Recycling of washed olive pomace ash for fired clay brick manufacturing. Constr Build Mater 61:320–326

Debbarma S, Ransinchung GD, Singh S, Sahdeo SK (2020) Utilization of industrial and agricultural wastes for productions of sustainable roller compacted concrete pavement mixes containing reclaimed asphalt pavement aggregates. Resour Conser Recycl 152:104504

Demeyer A, Nkana JV, Verloo M (2001) Characteristics of wood ash and influence on soil properties and nutrient uptake: an overview. Biores Technol 77:287–295

Demis S, Tapali JG, Papadakis VG (2014) An investigation of the effectiveness of the utilization of biomass ashes as pozzolanic materials. Constr Build Mater 68:291–300

Deo SV (2015) Mix design approach for high 28 days' strength, high-volume, low-lime fly ash concrete. Road Mater Pavement Design 16:707–715

Deschner F, Winnefeld F, Lothenbach B, Seufert S, Schwesig P, Dittrich S, Goetz-Neunhoeffer F, Neubauer J (2012) Hydration of Portland cement with high replacement by siliceous fly ash. Cem Concr Res 42:1389–1400

Dey KP, Ghosh S, Naskar MK (2013) Organic template-free synthesis of ZSM-5 zeolite particles using rice husk ash as silica source. Ceram Int 39:2153–2157

Dimitriou I, Eriksson J, Adler A, Aronsson P, Verwijst T (2006) Fate of heavy metals after application of sewage sludge and wood–ash mixtures to short-rotation willow coppice. Environ Pollut 142:160–169

Dindi A, Quang DV, Vega LF, Nashef E, Abu-Zahra MRM (2019) Applications of fly ash for CO_2 capture, utilization, and storage. J CO_2 Utili 29:82–102

Djamaluddin AR, Caronge MA, Tjaronge MW, Lando AT, Irmawaty R (2020) Evaluation of sustainable concrete paving blocks incorporating processed waste tea ash. Case Stud Constr Mater 12:e00325

Dwivedi S, Tripathi R, Srivastava S, Mishra S, Shukla M, Tiwari K, Singh R, Rai U (2007) Growth performance and biochemical responses of three rice (Oryza sativa L.) cultivars grown in fly-ash amended soil. Chemosphere 67:140–151

Eliche-Quesada D, Felipe-Sesé MA, Martínez-Martínez S, Pérez-Villarejo L (2017) Comparative study of the use of different biomass bottom ash in the manufacture of ceramic bricks. J Mater Civ Eng 29:04017238

Elliot AD, Zhang D-K (2005) Controlled release zeolite fertilisers: a value added product produced from fly ash. In: International ash utilization symposium (IAUS) and world of coal ash (WOCA), pp 1–32

Elliott L, Tittemore D, Papendick R, Cochran V, Bezdicek D (1982) The effect of mount st. Helens' ash on soil microbial respiration and numbers. J Environ Qual 11:164–166

Elseewi AA, Straughan I, Page A (1980) Sequential cropping of fly ash-amended soils: effects on soil chemical properties and yield and elemental composition of plants. Sci Total Environ 15:247–259

EPA US (2005) Using coal ash in highway construction: a guide to benefits and impacts. US EPA Washington, DC, USA

Etiegni L, Campbell AG, Mahler RL (1990) Wood ash as soil additive and liming agent. In: Environmental conference, TAPPI proceedings, Seattle, WA, 9–11 April 1990. Tappi Press, Atlanta, GA, pp 225–230

Etiegni L, Campbell A, Mahler R (1991) Evaluation of wood ash disposal on agricultural land. I. Potential as a soil additive and liming agent. Commun Soil Sci Plant Anal 22:243–256

Fail JL, Wochok ZS (1977) Soybean growth on fly ashamended strip mine spoils. Plant Soil 48:473–484

Fairbairn EMR, Americano BB, Cordeiro GC, Paula TP, Toledo Filho RD, Silvoso MM (2010) Cement replacement by sugar cane bagasse ash: CO_2 emissions reduction and potential for carbon credits. J Environ Manag 91:1864–1871

Fall M, Pokharel M (2010) Coupled effects of sulphate and temperature on the strength development of cemented tailings backfills: Portland cement-paste backfill. Cem Concr Compos 32:819–828

Fan Y, Weng Q, Zhuo Y, Dong S, Hu P, Li D (2019) Theoretical Study of As_2O_3 Adsorption Mechanisms on CaO surface. Materials 12:677

Fernández-Delgado Juárez M, Fabiani G, Mazzier T, Schönegger D, Pietramellara G, Gómez-Brandón M, Insam H (2019) Reclamation of acid soils with biomass ashes from pyrolytic wood liquefaction. Waste Biomass Valoriz

Fernández-Pereira C, de la Casa JA, Gómez-Barea A, Arroyo F, Leiva C, Luna Y (2011) Application of biomass gasification fly ash for brick manufacturing. Fuel 90:220–232

Flores CG, Schneider H, Marcilio NR, Ferret L, Oliveira JCP (2017) Potassic zeolites from Brazilian coal ash for use as a fertilizer in agriculture. Waste Manag 70:263–271

Flores Y, Flores R, Gallegos AA (2008) Heterogeneous catalysis in the Fenton-type system reactive black 5/H_2O_2. J Mol Catal A Chem 281:184–191

Fu J, Wang C, Chen X, Huang Z, Miu J, Shi L, Wang S (2019) Preparation and characterization of attapulgite-fly ash-based binuclear slow-release fertilizer encapsulated by eudragit. Agron J 111:2411–2421

Gaind S, Gaur A (2002) Impact of fly ash and phosphate solubilising bacteria on soybean productivity. Biores Technol 85:313–315

Gaind S, Gaur A (2004) Evaluation of fly ash as a carrier for diazotrophs and phosphobacteria. Biores Technol 95:187–190

Gajić G, Djurdjević L, Kostić O, Jarić S, Mitrović M, Stevanović B, Pavlović P (2016) Assessment of the phytoremediation potential and an adaptive response of Festuca rubra L. sown on fly ash deposits: native grass has a pivotal role in ecorestoration management. Ecol Eng 93:250–261

Gao H, Hu J (2018) Preparation method for fly ash coated biomass charcoal based slow-release fertilizer. Patent

Gao Y, Jiang J, Meng Y, Aihemaiti A, Ju T, Chen X, Yan F (2020) A novel nickel catalyst supported on activated coal fly ash for syngas production via biogas dry reforming. Renew Energy 149:786–793

Ghasemi Z, Younesi H, Kazemian H (2011) Synthesis of nanozeolite sodalite from rice husk ash without organic additives. Can J Chem Eng 89:601–608

Ghodrati M, Sims J, Vasilas B (1995) Evaluation of fly ash as a soil amendment for the Atlantic Coastal Plain: I. Soil hydraulic properties and elemental leaching. Water Air Soil Pollut 81:349–361

Givianrad MH, Rabani M, Saber-Tehrani M, Aberoomand-Azar P, Sabzevari MH (2013) Preparation and characterization of nanocomposite, silica aerogel, activated carbon and its adsorption properties for Cd (II) ions from aqueous solution. J Saudi Chem Soc 17:329–335

Gjyli S, Korpa A, Tabanelli T, Trettin R, Cavani F, Belviso C (2019) Higher conversion rate of phenol alkylation with diethylcarbonate by using synthetic fly ash-based zeolites. Microporous Mesoporous Mater 284:434–442

Gollakota AR, Volli V, Shu C-M (2019) Progressive utilisation prospects of coal fly ash: a review. Sci Total Environ 672:951–989

Gómez-Rey MX, Madeira M, Coutinho J (2013) Soil C and N dynamics, nutrient leaching and fertility in a pine plantation amended with wood ash under Mediterranean climate. Eur J Forest Res 132:281–295

Goyal D, Kaur K, Garg R, Vijayan V, Nanda S, Nioding A, Khanna S, Ramamurthy V (2002) Industrial fly ash as a soil amendment agent for raising forestry plantations, EPD congress 2002 and fundamentals of advanced materials for energy conversion as held at the 2002 TMS annual meeting, pp 251–260

Graham UM, Thomas G (1999) Chemically active aerogels for hot gas clean-up in a IGCC process. ACS Div Fuel Chem Prepr 44:706–710

Grantham L, Yosim S (1974) Recovery of lithium carbonate. Google Patents

Grice A, Finn T, Smith P (1999) Innovative use of backfill to enable continuous longwall mining through old headings at homestead colliery-wambo mining. A. dapted Aust Coal Rev

Gu Y, Zhang Y, Lin L, Xu H, Orndorff W, Pan W-P (2015) Evaluation of elemental mercury adsorption by fly ash modified with ammonium bromide. J Therm Anal Calorim 119:1663–1672

Güneyisi E, Gesoğlu M, Altan İ, Öz HÖ (2015) Utilization of cold bonded fly ash lightweight fine aggregates as a partial substitution of natural fine aggregate in self-compacting mortars. Constr Build Mater 74:9–16

Gupta PK, Mahato A, Oraon P, Gupta GK, Maity S (2020) Coal fly ash-derived mesoporous SBA-15 as support material for production of liquid hydrocarbon through Fischer–Tropsch route. Asia-Pac J Chem Eng, e2471

Gurupira T, Jones CL, Howard A, Lockert C, Wandell T, Stencel JM (2001) New products from coal combustion ash: selective extraction of particles with density <2. In: International ash utilization symposium

Guzel Kaya G, Deveci H (2020) Modified silica xerogel derived from groundnut hull ash by alkylammonium salt for epoxy nanocomposites: synergistic effects on thermal stability and flame retardancy. Thermochim Acta 689:178637

Harper JR, Mbakwe I (2020) The effectiveness of coal fly ash in the amelioration of acid soils of the South African highveld: a comparison with conventional liming materials. S Afr J Plant Soil, 1–7

Hassan S, Aigbodion V (2014) Effect coal ash on some refractory properties of alumino-silicate (Kankara) clay for furnace lining. Egypt J Basic Appl Sci 1:107–114

He J, Shi L (2012) Modified flue gas desulfurization residue (MFGDR)–a new type of acidic soil ameliorant and its effect on rice planting. J Clean Prod 24:159–167

He P, Zhang X, Peng X, Wu J, Chen N, Ren J (2015) Enhancement using external magnetic field on mercury capture by fly ash. Fuel 162:211–214

Helinski M, Fahey M, Fourie A (2011) Behavior of cemented paste backfill in two mine stopes: measurements and modeling. J Geotech Geoenviron Eng 137:171–182

Hill M, Lamp C (1980) Use of pulverised fuel ash from Victorian brown coal as a source of nutrients for a pasture species. Aust J Exp Agric 20:377–384

Hollman G, Steenbruggen G, Janssen-Jurkovičová M (1999) A two-step process for the synthesis of zeolites from coal fly ash. Fuel 78:1225–1230

Huang S-C, Chang F-C, Lo S-L, Lee M-Y, Wang C-F, Lin J-D (2007) Production of lightweight aggregates from mining residues, heavy metal sludge, and incinerator fly ash. J Hazard Mater 144:52–58

IS (2013) IS 3812-2: specification for pulverized fuel ash, Part 2: for use as admixture in cement mortar and concrete. Bureau of Indian Standards, New Delhi, India

Jain D, Khatri C, Rani A (2011) Synthesis and characterization of novel solid base catalyst from fly ash. Fuel 90:2083–2088

Jambhulkar HP, Shaikh SMS, Kumar MS (2018) Fly ash toxicity, emerging issues and possible implications for its exploitation in agriculture; Indian scenario: a review. Chemosphere 213:333–344

James AK, Thring RW, Helle S, Ghuman HS (2012) Ash management review—applications of biomass bottom ash. Energies 5:3856–3873

Jastrow JD, Zimmerman CA, Dvorak AJ, Hinchman RR (1979) Comparison of lime and fly ash as amendment to acidic coalmine refuse: growth responses and trace element uptake of two grasses. Agronne National Laboratory, Argonne, IL, p 43

Jiang N, Zhao J, Sun X, Bai L, Wang C (2017) Use of fly-ash slurry in backfill grouting in coal mines. Heliyon 3:e00470–e00470

Jokinen HK, Kiikkilä O, Fritze H (2006) Exploring the mechanisms behind elevated microbial activity after wood ash application. Soil Biol Biochem 38:2285–2291

Joseph K, Francis F, Chacko J, Das P, Hebbar G (2013) Fly ash cenosphere waste formation in coal fired power plants and its application as a structural material–a review. Int J Eng Res Technol 2:1236–1260

Kamal J, Mishra UK (2019) Fly ash utilization in lightweight aggregates for sustainable construction. Lecture Notes in Civil Engineering, pp 23–32

Kang A-H, Shang K, Ye D-D, Wang Y-T, Wang H, Zhu Z-M, Liao W, Xu S-M, Wang Y-Z, Schiraldi DA (2017) Rejuvenated fly ash in poly (vinyl alcohol)-based composite aerogels with high fire safety and smoke suppression. Chem Eng J 327:992–999

Kannan V, Ganesan K (2014) Chloride and chemical resistance of self compacting concrete containing rice husk ash and metakaolin. Constr Build Mater 51:225–234

Karasawa A, Suda S, Naito H, Fujiwara H (2003) Application of fly ash to concrete paving block. In: Proceedings of the 7th international conference on concrete block paving (Pave Africa). ISBN Number: 0-958-46091-4

Karpagavalli S, Ramabadran R (1997) Effect of lignite fly ash on the growth and dry matter production (DMP) of soil borne pathogens. In: Souvenir and Abstract: National Seminar on Bio-Utilization of Fly Ash, 4–5 April 1997. Khallikote Autonomous College, Berhampur, Orissa, India, p 11

Kaya GG, Yilmaz E, Deveci H (2019) Synthesis of sustainable silica xerogels/aerogels using inexpensive steel slag and bean pod ash: a comparison study. Adv Powder Technol

Kayali O (2008) Fly ash lightweight aggregates in high performance concrete. Constr Build Mater 22:2393–2399
Khan MR, Singh WN (2001) Effects of soil application of fly ash on the fusarial wilt on tomato cultivars. Int J Pest Manag 47:293–297
Khanna P, Raison RJ, Falkiner RA (1994) Chemical properties of ash derived from Eucalyptus litter and its effects on forest soils. For Ecol Manag 66:107–125
Khatib JM (2008) Performance of self-compacting concrete containing fly ash. Constr Build Mater 22:1963–1971
Khatri C, Mishra MK, Rani A (2010) Synthesis and characterization of fly ash supported sulfated zirconia catalyst for benzylation reactions. Fuel Process Technol 91:1288–1295
Kim B-J, Back J-H, Kim Y-S (1997) Effect of fly ash on the yield of Chinese cabbage and chemical properties of soil. Korean J Soil Sci Fertil 30:161–167
Kirubakaran D, Joseravindraraj B (2018) Utilization of pelletized fly ash aggregate to replace the natural aggregate: a review. Int J Civ Eng Technol 9:147–154
Kizinievic O, Kizinievic V (2016) Utilisation of wood ash from biomass for the production of ceramic products. Constr Build Mater 127:264–273
Krisnandi Y, Yanti F, Murti S (2017) Synthesis of ZSM-5 zeolite from coal fly ash and rice husk: characterization and application for partial oxidation of methane to methanol. In: IOP conference series: materialsscience and engineering, p 012031
Kucharski J, Jastrzębska E, Wyszkowska J (2006) Contamination of soil with hard coal ash as modifier of physicochemical and biological properties of soil. Electron J Pol Agric Univ 9:35
Kumar KV, Patra D (2012) Alteration in yield and chemical composition of essential oil of Mentha piperita L. plant: effect of fly ash amendments and organic wastes. Ecol Eng 47:237–241
Kumar KV, Singh N, Behl H, Srivastava S (2008) Influence of plant growth promoting bacteria and its mutant on heavy metal toxicity in Brassica juncea grown in fly ash amended soil. Chemosphere 72:678–683
Kumar MS, Singh L (2017) Critical review on types of bricks Type 13: wood ash bricks. Int J Mech Product Eng 5:81–83
Kumar PP (2017) Fly ash utilisation in the manufacture of green paver blocks and tiles. MTech thesis. National Institute of Technology, Rourkela, India
Kumar R, Ashish DK, Najia L (2015) Properties of non conventional (fly ash) brick: an experimental study. Int J Eng Trends Tech 24:198–204
Kumar R, Patyal V, Lallotra B, Ashish DK (2014) Study of properties of light weight fly ash brick. J Eng Res App, 49–53
Kumpiene J, Lagerkvist A, Maurice C (2007) Stabilization of Pb-and Cu-contaminated soil using coal fly ash and peat. Environ Pollut 145:365–373
Lal J, Mishra B, Sarkar A (1996) Effect of flyash on soil microbial and enzymatic activity. J Indian Soc Soil Sci 44:77–80
Lam MN-T, Le D-H, Jaritngam S (2018) Compressive strength and durability properties of roller-compacted concrete pavement containing electric arc furnace slag aggregate and fly ash. Constr Build Mater 191:912–922
Lee H, Ha HS, Lee CH, Lee YB, Kim PJ (2006) Fly ash effect on improving soil properties and rice productivity in Korean paddy soils. Biores Technol 97:1490–1497
Li L, Aubertin M (2008) Influence of water pressure on the stress state in stopes with cohesionless backfill. Geotech Geol Eng 27:1
Li L, Wang S, Zhu Z, Yao X, Yan Z (2008) Catalytic decomposition of ammonia over fly ash supported Ru catalysts. Fuel Process Technol 89:1106–1112
Li Z-D, Wang H-L, Wei X-N, Liu X-Y, Yang Y-F, Jiang W-F (2016) Preparation and photocatalytic performance of magnetic Fe_3O_4@TiO_2 core–shell microspheres supported by silica aerogels from industrial fly ash. J Alloy Compd 659:240–247
Liiri M, Haimi J, Setälä H (2002) Community composition of soil microarthropods of acid forest soils as affected by wood ash application. Pedobiologia 46:108–124

Lin K-L, Chang J-C (2013) Feasibility of recycling waste diatomite and fly ash cosintered as porous ceramics. Environ Prog Sustain Energy 32:25–34

Liu P, Huang R, Tang Y (2019) Comprehensive Understandings of Rare Earth Element (REE) Speciation in Coal Fly Ashes and Implication for REE Extractability. Environ Sci Technol 53:5369–5377

Liu S, Yang C, Liu W, Yi L, Qin W (2020) A novel approach to preparing ultra-lightweight ceramsite with a large amount of fly ash. Front Environ Sci Eng 14:62

Loh YR, Sujan D, Rahman ME, Das CA (2013) Sugarcane bagasse—the future composite material: a literature review. Resour Conserv Recycl 75:14–22

Lopareva-Pohu A, Pourrut B, Waterlot C, Garçon G, Bidar G, Pruvot C, Shirali P, Douay F (2011) Assessment of fly ash-aided phytostabilisation of highly contaminated soils after an 8-year field trial: part 1. Influence on soil parameters and metal extractability. Sci Total Environ 409:647–654

Lu S-G, Sun F-F, Zong Y-T (2014) Effect of rice husk biochar and coal fly ash on some physical properties of expansive clayey soil (Vertisol). Catena 114:37–44

Lu S, Zhu L (2004) Effect of fly ash on physical properties of ultisols from subtropical China. Commun Soil Sci Plant Anal 35:703–717

Lu Y, Li X, Yin X, Utomo HD, Tao NF, Huang H (2018) Silica aerogel as super thermal and acoustic insulation materials. J Environ Prot 9:295–308

Lundkvist H (1998) Wood ash effects on enchytraeid and earthworm abundance and enchytraeid cadmium content. Scand J For Res 13:86–95

Luo Y, Zheng S, Ma S, Liu C, Wang X (2017) Ceramic tiles derived from coal fly ash: preparation and mechanical characterization. Ceram Int 43:11953–11966

Lura P, Wyrzykowski M, Tang C, Lehmann E (2014) Internal curing with lightweight aggregate produced from biomass-derived waste. Cem Concr Res 59:24–33

Ma L, Han L, Chen S, Hu J, Chang L, Bao W, Wang J (2019a) Rapid synthesis of magnetic zeolite materials from fly ash and iron-containing wastes using supercritical water for elemental mercury removal from flue gas. Fuel Process Technol 189:39–48

Ma W, Zong P, Cheng Z, Wang B, Sun Q (2014) Adsorption and bio-sorption of nickel ions and reuse for 2-chlorophenol catalytic ozonation oxidation degradation from water. J Hazard Mater 266:19–25

Ma Z, Shan X, Cheng F (2019b) Distribution characteristics of valuable elements, Al, Li, and Ga, and rare earth elements in feed coal, fly ash, and bottom ash from a 300 MW circulating fluidized bed boiler. ACS Omega 4:6854–6863

Maitra S (1999) Ceramic products from fly ash: global perspectives. In: Chatterjee B, Singh KK, Goswami NG (eds) Proceeding of fly ash utilisation for value added products'. CSIR-NML, Jamshedpur, pp 32–37

Malhotra VM, Mehta PK (2002) High-performance, high-volume fly ash concrete: materials, mixture proportioning, properties, construction practice, and case histories. Canada, p 101. (ISBN 0-9731507-0-X; TRN: 000600211)

Manoharan V, Yunusa I, Loganathan P, Lawrie R, Skilbeck C, Burchett M, Murray B, Eamus D (2010) Assessments of Class F fly ashes for amelioration of soil acidity and their influence on growth and uptake of Mo and Se by canola. Fuel 89:3498–3504

Mardani-Aghabaglou A, Ramyar K (2013) Mechanical properties of high-volume fly ash roller compacted concrete designed by maximum density method. Constr Build Mater 38:356–364

Marinković S, Dragaš J (2018) 11-fly ash. In: Siddique R, Cachim P (eds) Waste and supplementary cementitious materials in concrete. Woodhead Publishing, pp 325–360

Mario C-M (2006) Industrial ecology approach to management of fly ash from fluidized bed combustion: production of slow-release fertilizer and soil conditioner. Theses & Dissertations. University of Puerto Rico, Mayaguez Campus. https://hdl.handle.net/20.500.11801/1781

Martens D (1971) Availability of plant nutrients in fly ash. Compost Sci United States) 12

Masto RE, Mahato M, Selvi VA, Ram LC (2013) The effect of fly ash application on phosphorus availability in an acid soil. Energy Sour Part A Recover Utili Environ Eff 35:2274–2283

Masued GG (2017) Effect of sugar cane bagasse ash on the physical properties of subgrade layer. J Eng Sustain Dev 21:146–158
Matsi T, Keramidas V (1999) Fly ash application on two acid soils and its effect on soil salinity, pH, B, P and on ryegrass growth and composition. Environ Pollut 104:107–112
McCarty G, Siddaramappa R, Wright R, Codling E, Gao G (1994) Evaluation of coal combustion byproducts as soil liming materials: their influence on soil pH and enzyme activities
Mehta P, Monteiro PJM (2006) Concrete: microstructure, properties and materials, 3rd edn. McGraw-Hill, New York
Mehta PK (2004) High-performance, high-volume fly ash concrete for sustainable development. In: Proceedings of the international workshop on sustainable development and concrete technology. Iowa State University Ames, IA, USA, pp 3–14
Mercl F, Tejnecký V, Száková J, Tlustoš P (2016) Nutrient dynamics in soil solution and wheat response after biomass ash amendments. Agron J 108:2222–2234
Mishra DP, Das SK (2010) A study of physico-chemical and mineralogical properties of Talcher coal fly ash for stowing in underground coal mines. Mater Charact 61:1252–1259
Mittra B, Karmakar S, Swain D, Ghosh B (2005) Fly ash—a potential source of soil amendment and a component of integrated plant nutrient supply system. Fuel 84:1447–1451
Mohamed R, Mkhalid I, Barakat M (2015) Rice husk ash as a renewable source for the production of zeolite NaY and its characterization. Arab J Chem 8:48–53
Moisés MP, da Silva CTP, Meneguin JG, Girotto EM, Radovanovic E (2013) Synthesis of zeolite NaA from sugarcane bagasse ash. Mater Lett 108:243–246
Molina M, Fuentes R, Calderón R, Escudey M, Avendaño K, Gutiérrez M, Chang AC (2007) Impact of forest fire ash on surface charge characteristics of Andisols. Soil Sci 172:820–834
Mondal S, Ghar A, Satpati AK, Sinharoy P, Singh DK, Sharma JN, Sreenivas T, Kain V (2019) Recovery of rare earth elements from coal fly ash using TEHDGA impregnated resin. Hydrometallurgy 185:93–101
Moreno-Maroto JM, Camacho PN, Cotes-Palomino T, García CM, Alonso-Azcárate J (2019) Manufacturing of lightweight aggregates from biomass fly ash, beer bagasse, Zn-rich industrial sludge and clay by slow firing. J Environ Manag 246:785–795
Moriyama R, Takeda S, Onozaki M, Katayama Y, Shiota K, Fukuda T, Sugihara H, Tani Y (2005) Large-scale synthesis of artificial zeolite from coal fly ash with a small charge of alkaline solution. Fuel 84:1455–1461
Munda S, Nayak AK, Mishra PN, Bhattacharyya P, Mohanty S, Kumar A, Kumar U, Baig MJ, Tripathi R, Shahid M, Adak T, Thilagam VK (2016) Combined application of rice husk biochar and fly ash improved the yield of lowland rice. Soil Res 54:451–459
Mupambwa HA, Dube E, Mnkeni PN (2015) Fly ash composting to improve fertiliser value-a review. S Afr J Sci 111:1–6
Mushtaq F, Zahid M, Bhatti IA, Nasir S, Hussain T (2019) Possible applications of coal fly ash in wastewater treatment. J Environ Manag 240:27–46
Naskar MK, Kundu D, Chatterjee M (2012) A facile hydrothermal conversion of rice husk ash to ZSM-5 zeolite powders. J Am Ceram Soc 95:925–930
Nazriati N, Setyawan H, Affandi S, Yuwana M, Winardi S (2014) Using bagasse ash as a silica source when preparing silica aerogels via ambient pressure drying. J Non-Cryst Solids 400:6–11
Ng E-P, Awala H, Tan K-H, Adam F, Retoux R, Mintova S (2015) EMT-type zeolite nanocrystals synthesized from rice husk. Microporous Mesoporous Mater 204:204–209
Ngu L-N, Wu H, Zhang D-K (2007) Characterization of ash cenospheres in fly ash from Australian power stations. Energy Fuels 21:3437–3445
NTPC (2007) Fly ash for cement concrete. Resource for high strength and durability of structures at lower cost. Ash Utilization Division, NTPC Limited, A-11, NFL Premises, Sector–24, Noida, India
NTPC (2020) Fly ash bricks. A useful and environmental friendly building product. Ash Utilisation Division. National Thermap Power Corporation Limited (NTPC). https://www.ntpc.co.in/ash-download/1675/0/fly-ash-bricks. Accessed on 15 May 2020.

Ochecova P, Tlustos P, Szakova J (2014) Wheat and soil response to wood fly ash application in contaminated soils. Agron J 106:995–1002
Ochecova P, Tlustos P, Szakova J, Mercl F, Maciak M (2016) Changes in nutrient plant availability in loam and sandy clay loam soils after wood fly and bottom ash amendment. Agron J 108:487–497
Ochedi FO, Liu Y, Hussain A (2020) A review on coal fly ash-based adsorbents for mercury and arsenic removal. J Clean Prod, 122143
Otero JG, Blanco F, Garcia M, Ayala J (2004) Manufacture of refractory insulating bricks using fly ash and clay. Br Ceram Trans 103:181–186
Ozolincius R, Armolaitis K, Raguotis A, Varnagiryte I, Zenkovaite J (2006) Influence of wood ash recycling on chemical and biological condition of forest Arenosols. J For Sci 52:79–86
Page A, Elseewi AA, Straughan I (1979) Physical and chemical properties of fly ash from coal-fired power plants with reference to environmental impacts, residue reviews. Springer, pp 83–120
Palumbo A, Mccarthy JF, Amonette JE, Fisher L, Wullschleger SD, Daniels WL (2004) Prospects for enhancing carbon sequestration and reclamation of degraded lands with fossil-fuel combustion by-products. Adv Environ Res 8:425–438
Pan J, Zhou C, Liu C, Tang M, Cao S, Hu T, Ji W, Luo Y, Wen M, Zhang N (2018) Modes of occurrence of rare earth elements in coal fly ash: a case study. Energy Fuels 32:9738–9743
Pandey VC (2020a) Chapter 3-Scope of fly ash use in agriculture: prospects and challenges. In: Pandey VC (ed) Phytomanagement of fly ash. Elsevier, pp 63–101
Pandey VC (2020b) Chapter 7-Afforestation on fly ash catena: an adaptive fly ash management. In: Pandey VC (ed) Phytomanagement of fly ash. Elsevier, pp 195–234
Pandey VC, Abhilash P, Upadhyay RN, Tewari D (2009a) Application of fly ash on the growth performance and translocation of toxic heavy metals within Cajanus cajan L.: implication for safe utilization of fly ash for agricultural production. J Hazard Mater 166:255–259
Pandey VC, Abhilash PC, Singh N (2009b) The Indian perspective of utilizing fly ash in phytoremediation, phytomanagement and biomass production. J Environ Manage 90:2943–2958
Pandey VC, Singh K, Singh JS, Kumar A, Singh B, Singh RP (2012a) Jatropha curcas: a potential biofuel plant for sustainable environmental development. Renew Sustain Energy Rev 16:2870–2883
Pandey VC, Singh K, Singh RP, Singh B (2012b) Naturally growing Saccharum munja L. on the fly ash lagoons: a potential ecological engineer for the revegetation and stabilization. Ecol Eng 40:95–99
Pandey VC, Singh N (2010) Impact of fly ash incorporation in soil systems. Agr Ecosyst Environ 136:16–27
Park D, Middleton A, Smith R, Deblonde G, Laudal D, Theaker N, Hsu-Kim H, Jiao Y (2020) A biosorption-based approach for selective extraction of rare earth elements from coal byproducts. Sep Purif Technol 241:116726
Park ND, Rutherford PM, Thring RW, Helle SS (2012) Wood pellet fly ash and bottom ash as an effective liming agent and nutrient source for rye grass (Lolium perenne L.) and oats (Avena sativa). Chemosphere 86:427–432
Parkinson D, Coleman DC (1991) Microbial communities, activity and biomass. Agr Ecosyst Environ 34:3–33
Pathan S, Aylmore L, Colmer T (2003) Soil properties and turf growth on a sandy soil amended with fly ash. Plant Soil 256:103–114
Patterson SJ, Acharya SN, Thomas JE, Bertschi AB, Rothwell RL (2004) Barley biomass and grain yield and canola seed yield response to land application of wood ash. Agron J 96:971–977
Pattnaik T, Mohanta N, Patel N, Biswal P, Moharana R (2018) Manufacture of interlocking concrete paving blocks with fly ash and glass powder. Int J Eng Sci Res Technol 7(1):604–612
Payá J, Monzó J, Borrachero MV, Tashima MM, Soriano Z (2018) Waste and supplementary cementitious materials in concrete. Characterisation, properties and applications. Woodhead Publishing Series in Civil and Structural Engineering, pp 559–598
Pedišius N, Kulokas M, Praspaliauskas M, Zaleskas V, Buinevičius K (2019) Properties of biofuel fly ash and capabilities of its use for agricultural needs

Pérez-Villarejo L, Eliche-Quesada D, Carrasco-Hurtado B, Sánchez-Soto PJ (2020) Valorization of olive biomass fly ash for production eco friendly ceramic bricks. In: Hashmi S, Choudhury IA (eds) Encyclopedia of renewable and sustainable materials. Elsevier, Oxford, pp 285–294

Pérez-Villarejo L, Eliche-Quesada D, Iglesias-Godino FJ, Martínez-García C, Corpas-Iglesias FA (2012) Recycling of ash from biomass incinerator in clay matrix to produce ceramic bricks. J Environ Manag 95:S349–S354

Phung H, Lam H, Page A, Lund L (1979) The practice of leaching boron and soluble salts from fly ash-amended soils. Water Air Soil Pollut 12:247–254

Pichtel J, Hayes J (1990) Influence of fly ash on soil microbial activity and populations. J Environ Qual 19:593–597

Prashant, Ghosh CN, Mandal PK, Kumar V, Sinha A (2005) Paste fill–the future of filling in underground coal mines. Fly Ash India 2005, New Delhi

Purnomo CW, Salim C, Hinode H (2012) Synthesis of pure Na–X and Na–A zeolite from bagasse fly ash. Microporous Mesoporous Mater 162:6–13

Qi L, Sun S, Zhang Y, Zhu Z, Wu B (2019) Environment-friendly type fly ash coated release-controlled fertilizer and preparation method. Patent CN109232090

Qiu Q, Hlavacek V (2010) Energy estimation on CRN process of fly ash as a slow-release nitrogen fertilizer. Ind Eng Chem Res 49:5939–5944

Quirantes M, Calvo F, Romero E, Nogales R (2016) Soil-nutrient availability affected by different biomass-ash applications. J Soil Sci Plant Nutr 16:159–163

Rahmawati NUS, Sutopo NR, Nuraini Y, Handayanto E (2020) Phytotoxicity of coal fly ash on plant growth and heavy metal uptake by plant in an acid soil. J Degrad Min Lands Manag 7:2233–2240

Ram L, Masto R (2014) Fly ash for soil amelioration: a review on the influence of ash blending with inorganic and organic amendments. Earth Sci Rev 128:52–74

Ram L, Srivastava N, Tripathi R, Jha S, Sinha AK, Singh G, Manoharan V (2006) Management of mine spoil for crop productivity with lignite fly ash and biological amendments. J Environ Manag 79:173–187

Ram LC, Masto RE (2010) An appraisal of the potential use of fly ash for reclaiming coal mine spoil. J Environ Manag 91:603–617

Ram LC, Srivastava NK, Jha SK, Sinha AK, Masto RE, Selvi VA (2007) Management of lignite fly ash for improving soil fertility and crop productivity. Environ Manag 40:438–452

Rani R, Jain MK (2017) Effect of bottom ash at different ratios on hydraulic transportation of fly ash during mine fill. Powder Technol 315:309–317

Ranjbar N, Kuenzel C (2017) Cenospheres: a review. Fuel 207:1–12

Rao B, Kumar V (2019) Use of fly ash in mining sector. In: Ghosh S (ed) Waste valorisation and recycling. Springer, Singapore

Rashad AM (2015) A brief on high-volume Class F fly ash as cement replacement–a guide for civil engineer. Int J Sustain Built Environ 4:278–306

Rau N, Mishra V, Sharma M, Das MK, Ahaluwalia K, Sharma RS (2009) Evaluation of functional diversity in rhizobacterial taxa of a wild grass (Saccharum ravennae) colonizing abandoned fly ash dumps in Delhi urban ecosystem. Soil Biol Biochem 41:813–821

Reddy A, Gupta V, Garg D (2015) Effect of partial replacement of cement by rice husk ash using nylon fiber in concrete paver block. IJSRD Int J Sci Res Dev. ISSN (online) 2321-0613

Rivera F, Martínez P, Castro J, López M (2015) Massive volume fly-ash concrete: a more sustainable material with fly ash replacing cement and aggregates. Cem Concr Compos 63:104–112

Rosita W, Bendiyasa IM, Perdana I, Anggara F (2020) Sequential particle-size and magnetic separation for enrichment of rare-earth elements and yttrium in Indonesia coal fly ash. J Environ Chem Eng 8:103575

Rout K, Pal A, Das I, Mandal M, Nayak R, Jena D, Satpathy M (2016) Characterization of fly ash and its use in sesamum on an acid soil. J Indian Soc Soil Sci 64:86–92

Rukzon S, Chindaprasirt P, Mahachai R (2009) Effect of grinding on chemical and physical properties of rice husk ash. Int J Miner Metall Mater 16:242–247

Rybak W, Moroń W, Czajka KM, Kisiela AM, Ferens W, Jodkowski W, Andryjowicz C (2017) Co-combustion of unburned carbon separated from lignite fly ash. Energy Procedia 120:197–205
Sahoo PK, Kim K, Powell M, Equeenuddin SM (2016) Recovery of metals and other beneficial products from coal fly ash: a sustainable approach for fly ash management. Int J Coal Sci Technol 3:267–283
Saputra E, Muhammad S, Sun H, Ang HM, Tadé MO, Wang S (2012) Red mud and fly ash supported Co catalysts for phenol oxidation. Catal Today 190:68–72
Sarangi PK, Mahakur D, Mishra PC (2001) Soil biochemical activity and growth response of rice Oryza sativa in flyash amended soil. Biores Technol 76:199–205
Saucedo-Salazar EM, Perera-Mercado YA, Rodríguez-Ruiz FR, Arauza-Villarreal A (2014) Production of glass spheres from blast furnace slags by a thermal flame projection process. Ceram Int 40:1177–1182
Schiemenz K, Kern J, Paulsen HM, Bachmann S, Eichler-Löbermann B (2011a) Phosphorus fertilizing effects of biomass ashes. In: Insam H, Knapp BA (eds) Recycling of biomass ashes. Springer, Berlin, Germany, pp 17–31
Schiemenz K, Kern J, Paulsen H-M, Bachmann S, Eichler-Löbermann B (2011b) Phosphorus fertilizing effects of biomass ashes, recycling of biomass ashes. Springer, pp 17–31
Schnappinger MG Jr, Martens DC, Plank CO (1975) Zinc availability as influenced by application of fly ash to soil. Environ Sci Technol 9:258–261
Schutter M, Fuhrmann J (2001) Soil microbial community responses to fly ash amendment as revealed by analyses of whole soils and bacterial isolates. Soil Biol Biochem 33:1947–1958
Scotti IA, Silva S, Botteschi G (1999) Effect of fly ash on the availability of Zn, Cu, Ni and Cd to chicory. Agr Ecosyst Environ 72:159–163
Segawa H, Akizuki K (1981) Apparatus for producing potassium silicate fertilizer. Google Patents
Seleiman MF, Kheir AMS (2018) Saline soil properties, quality and productivity of wheat grown with bagasse ash and thiourea in different climatic zones. Chemosphere 193:538–546
Sen A (1997) Microbial population dynamics in fly ash amended acid lateritic soil. B. Tech. Thesis. Indian Institute of Technology, Kharagpur, India
Seneviratne SI, Corti T, Davin EL, Hirschi M, Jaeger EB, Lehner I, Orlowsky B, Teuling AJ (2010) Investigating soil moisture–climate interactions in a changing climate: a review. Earth Sci Rev 99:125–161
Seoane S, Leiros M (2001) Acidification–neutralization processes in a lignite mine spoil amended with fly ash or limestone. J Environ Qual 30:1420–1431
Shah BA, Patel HD, Shah AV (2011) Equilibrium and kinetic studies of the adsorption of basic dye from aqueous solutions by zeolite synthesized from bagasse fly ash. Environ Prog Sustain Energy 30:549–557
Shaheen S, Tsadilas C (2010) Sorption of cadmium and lead by acidic Alfisols as influenced by fly ash and sewage sludge application. Pedosphere 20:436–445
Shaheen SM, Hooda PS, Tsadilas CD (2014) Opportunities and challenges in the use of coal fly ash for soil improvements–a review. J Environ Manag 145:249–267
Sharma SK, Kalra N (2006) Effect of flyash incorporation on soil properties and productivity of crops: a review
Shi M, Luo G, Zhu H, Zou R, Hu J, Xu Y, Yao H (2019) Surface modification of fly ash by non-thermal air plasma for elemental mercury removal from coal-fired flue gas. Environ Technol, 1–12
Shi R-Y, Li J-Y, Xu R-K, Qian W (2016) Ameliorating effects of individual and combined application of biomass ash, bone meal and alkaline slag on acid soils. Soil Tillage Res 162:41–45
Shi R, Li J, Jiang J, Mehmood K, Liu Y, Xu R, Qian W (2017) Characteristics of biomass ashes from different materials and their ameliorative effects on acid soils. J Environ Sci 55:294–302
Shin J-S, Seong K-S, Choi D-H (1994) Utilization of fly ash as a source of mineral fertilizers-IV. Development of slowly released K fertilizer. Korean J Soil Sci Fertil 27:10–14
Sikka R, Kansal B (1995) Effect of fly-ash application on yield and nutrient composition of rice, wheat and on pH and available nutrient status of soils. Biores Technol 51:199–203

Singh A, Agrawal S (2010) Response of mung bean cultivars to fly ash: growth and yield. Ecotoxicol Environ Saf 73:1950–1958
Singh A, Sharma RK, Agrawal SB (2008) Effects of fly ash incorporation on heavy metal accumulation, growth and yield responses of Beta vulgaris plants. Biores Technol 99:7200–7207
Singh B (2018) Rice husk ash. In: Waste and supplementary cementitious materials in concrete characterisation, properties and applications. Woodhead Publishing Series in Civil and Structural Engineering, pp 417–460
Singh LP, Siddiqui ZA (2003) Effects of fly ash and Helminthosporium oryzae on growth and yield of three cultivars of rice. Biores Technol 86:73–78
Singh N, Kumar P, Goyal P (2019) Reviewing the behaviour of high volume fly ash based self compacting concrete. J Build Eng 26:100882
Sinha S, Gupta AK (2005) Translocation of metals from fly ash amended soil in the plant of Sesbania cannabina L. Ritz: effect on antioxidants. Chemosphere 61:1204–1214
Sivakugan N, Rankine R, Rankine K, Rankine K (2006) Geotechnical considerations in mine backfilling in Australia. J Clean Prod 14:1168–1175
Sklivaniti V, Tsakiridis PE, Katsiotis NS, Velissariou D, Pistofidis N, Papageorgiou D, Beazi M (2017) Valorisation of woody biomass bottom ash in Portland cement: a characterization and hydration study. J Environ Chem Engi 5:205–213
Smirnova I, Mamic J, Arlt W (2003) Adsorption of drugs on silica aerogels. Langmuir 19:8521–8525
Song M, Lin S, Takahashi F (2020) Coal fly ash amendment to mitigate soil water evaporation in arid/semi-arid area: an approach using simple drying focusing on sieve size and temperature. Resour Conserv Recycl 156:104726
Souza AE, Teixeira SR, Santos GTA, Costa FB, Longo E (2011) Reuse of sugarcane bagasse ash (SCBA) to produce ceramic materials. J Environ Manage 92:2774–2780
Srivastava A, Chhonkar P (2000) Amelioration of coal mine spoils through fly ash application as liming material
Sua-iam G, Makul N (2013a) Use of increasing amounts of bagasse ash waste to produce self-compacting concrete by adding limestone powder waste. J Clean Prod 57:308–319
Sua-iam G, Makul N (2013b) Utilization of limestone powder to improve the properties of self-compacting concrete incorporating high volumes of untreated rice husk ash as fine aggregate. Constr Build Mater 38:455–464
Sua-Iam G, Makul N (2015) Utilization of coal- and biomass-fired ash in the production of self-consolidating concrete: a literature review. J Clean Prod 100:59–76
Suhas B, Mahapatra DR, Acharaylu VVN (2019) An granulation method for granulating sugar distillery ash into fertilizer granules. Indian Patent 987/CHE/2013
Sukkae R, Suebthawilkul S, Cherdhirunkorn B (2018) Utilization of coal fly ash as a raw material for refractory production. J Met Mater Miner, 28
Sun J, Shen X, Tan G, Tanner JE (2019) Compressive strength and hydration characteristics of high-volume fly ash concrete prepared from fly ash. J Therm Anal Calorim 136:565–580
Taggart RK, Hower JC, Dwyer GS, Hsu-Kim H (2016) Trends in the rare earth element content of U.S.-based coal combustion fly ashes. Environ Sci Technol 50:5919–5926
Tarkalson D, Hergert G, Stevens W, McCallister D, Kachman SD (2005) Fly ash as a liming material for corn PRODUCTION1. Soil Sci 170:386–398
Tolle DA, Arthur MF, Van Voris P (1983) Microcosm/field comparison of trace element uptake in crops grown in fly ash-amended soil. Sci Total Environ 31:243–261
Tripathi RC, Masto RE, Ram LC (2009) Bulk use of pond ash for cultivation of wheat–maize–eggplant crops in sequence on a fallow land. Resour Conserv Recycl 54:134–139
Tsadilas C, Shaheen S, Samaras V (2009) Influence of coal fly ash application individually and mixing with sewage sludge on wheat growth and soil chemical properties under field conditions. In: 15th international symposium on environmental pollution and its impact on life in the mediterranean region, October, 7e11, Bari, Italy, p 145

Twardowska I (1999) Environmental aspects of power plants fly ash utilization in deep coal mine workings. In: Biogeochemistry of trace elements in coal and coal combustion byproducts. Springer, pp 29–57
Usanova K (2020) Properties of cold-bonded fly ash lightweight aggregate concretes. Lecture Notes in Civil Engineering, pp 507–516
Vallini G, Vaccari F, Pera A, Agnolucci M, Scatena S, Varallo G (1999) Evaluation of cocomposted coal fly ash on dynamics of microbial populations and heavy metal uptake. Compos Sci Util 7:81–90
Vassilev SV, Baxter D, Andersen LK, Vassileva CG (2013a) An overview of the composition and application of biomass ash. Part 1. Phase–mineral and chemical composition and classification. Fuel 105:40–76
Vassilev SV, Baxter D, Andersen LK, Vassileva CG (2013b) An overview of the composition and application of biomass ash.: Part 2. Potential utilisation, technological and ecological advantages and challenges. Fuel 105:19–39
Vassilev SV, Vassileva CG (2020) Contents and associations of rare earth elements and yttrium in biomass ashes. Fuel 262:116525
Visa M, Isac L, Duta A (2012) Fly ash adsorbents for multi-cation wastewater treatment. Appl Surf Sci 258:6345–6352
Vories KC (2005) Placement of coal combustion by products at Surface Mining Control and Reclamation Act (SMCRA) mines: a short history of OSM technical efforts and responses to environmental concerns. In: Proceedings of national meeting of the American society of mining and reclamation
Wang L, Zhang G, Wang P, Yu S (2018a) Effects of fly ash and crystalline additive on mechanical properties of two-graded roller compacted concrete in a high RCC arch dam. Constr Build Mater 182:682–690
Wang N, Zhao Q, Li Q, Zhang G, Huang Y (2020) Degradation of polyacrylamide in an ultrasonic-Fenton-like process using an acid-modified coal fly ash catalyst. Powder Technol 369:270–278
Wang S, Lu G (2007) Effect of chemical treatment on Ni/fly-ash catalysts in methane reforming with carbon dioxide. In: Studies in surface science and catalysis. Elsevier, pp 275–280
Wang S, Miller A, Llamazos E, Fonseca F, Baxter L (2008) Biomass fly ash in concrete: mixture proportioning and mechanical properties. Fuel 87:365–371
Wang S, Zhang Y, Gu Y, Wang J, Zhang Y, Cao Y, Romero CE, Pan W-P (2016) Using modified fly ash for mercury emissions control for coal-fired power plant applications in China. Fuel 181:1230–1237
Wang X-M, Bin Z, Zhang C-S, Zhang Q-L (2009) Paste-like self-flowing transportation backfilling technology based on coal gangue. Min Sci Technol (China) 19:137–143
Wang X, Chen A, Liu H (2019) Method for preparing silicon fertilizer from fly ash. Patent CN110272299
Wang Y, Du T, Jia H, Qiu Z, Song Y (2018b) Synthesis, characterization and CO_2 adsorption of NaA, NaX and NaZSM-5 from rice husk ash. Solid State Sci 86:24–33
Wang Y, Du T, Qiu Z, Song Y, Che S, Fang X (2018c) CO_2 adsorption on polyethylenimine-modified ZSM-5 zeolite synthesized from rice husk ash. Mater Chem Phys 207:105–113
Wong J, Lai K (1996) Effect of an artificial soil mix from coal fly ash and sewage sludge on soil microbial activity. Biol Fertil Soils 23:420–424
Wong JW, Fang M, Jiang R (2001) Persistency of bacterial indicators in biosolids stabilization with coal fly ash and lime. Water Environ Res 73:607–611
Wong MH, Wong J (1986) Effects of fly ash on soil microbial activity. Environ Pollut Ser A Ecol Biol 40:127–144
Wu X, Fan M, McLaughlin JF, Shen X, Tan G (2018) A novel low-cost method of silica aerogel fabrication using fly ash and trona ore with ambient pressure drying technique. Powder Technol 323:310–322
Xing L, Xu Y, Zhong Q (2012) Mn and Fe modified fly ash as a superior catalyst for elemental mercury capture under air conditions. Energy Fuels 26:4903–4909

Xu Y, Zhong Q, Xing L (2014) Gas-phase elemental mercury removal from flue gas by cobalt-modified fly ash at low temperatures. Environ Technol 35:2870–2877

Xuan X, Yue C, Li S, Yao Q (2003) Selective catalytic reduction of NO by ammonia with fly ash catalyst☆. Fuel 82:575–579

Yang J, Zhao Y, Zhang J, Zheng C (2016) Removal of elemental mercury from flue gas by recyclable $CuCl_2$ modified magnetospheres catalyst from fly ash. Part 1. Catalyst characterization and performance evaluation. Fuel 164:419–428

Yang J, Zhao Y, Zhang S, Liu H, Chang L, Ma S, Zhang J, Zheng C (2017) Mercury removal from flue gas by magnetospheres present in fly ash: role of iron species and modification by HF. Fuel Process Technol 167:263–270

Yang Y, Xue QH, Li SD (2012) Study on preparation of sintered thermal insulation wall material by high addition fly ash. In: Advanced Materials Research. Trans Tech Publication, pp 1258–1264

Yao Z, Ji X, Sarker P, Tang J, Ge L, Xia M, Xi Y (2015) A comprehensive review on the applications of coal fly ash. Earth Sci Rev 141:105–121

Yao ZT, Xia MS, Ye Y, Zhang L (2009) Synthesis of zeolite Li-ABW from fly ash by fusion method. J Hazard Mater 170:639–644

Yeledhalli N, Prakash S, Gurumurthy S, Ravi M (2010) Coal fly ash as modifier of physico-chemical and biological properties of soil. Karnataka J Agric Sci, 20

Yoga KE, Febrina EG, Haqiqie FA (2019) The usage of coal combustion (fly) ash as mitigation method to control soil erosion and water pollution: a state knowledge review. In: Journal of Physics: Conference Series. IOP Publishing, p 012036

Yu C-L, Deng Q, Jian S, Li J, Dzantor EK, Hui D (2019) Effects of fly ash application on plant biomass and element accumulations: a meta-analysis. Environ Pollut 250:137–142

Yu Y-T (2004) Preparation of nanocrystalline TiO_2-coated coal fly ash and effect of iron oxides in coal fly ash on photocatalytic activity. Powder Technol 146:154–159

Yunusa I, Eamus D, DeSilva D, Murray B, Burchett M, Skilbeck G, Heidrich C (2006) Fly-ash: an exploitable resource for management of Australian agricultural soils. Fuel 85:2337–2344

Zeng X, Xu Y, Zhang B, Luo G, Sun P, Zou R, Yao H (2017) Elemental mercury adsorption and regeneration performance of sorbents $FeMnO_x$ enhanced via non-thermal plasma. Chem Eng J 309:503–512

Zhang K, Wang P, Zhang D, Zhang K (2019) Studies on the correlation between physicochemical properties of fly ash and its sorption of gas-phase arsenic. Environ Technol 40:2548–2555

Zhang K, Zhang D, Zhang K, Cao Y (2016a) Capture of gas-phase arsenic by ferrospheres separated from fly ashes. Energy Fuels 30:8746–8752

Zhang R, Feng J, Cheng X, Gong L, Li Y, Zhang H (2014a) Porous thermal insulation materials derived from fly ash using a foaming and slip casting method. Energy Build 81:262–267

Zhang Y, Duan W, Liu Z, Cao Y (2014b) Effects of modified fly ash on mercury adsorption ability in an entrained-flow reactor. Fuel 128:274–280

Zhang Y, Wang C, Liu H (2016b) Experiment and mechanism research on gas-phase As_2O_3 adsorption of $Fe_2O_3/\gamma\text{-}Al_2O_3$. Fuel 181:1034–1040

Zhang Z, Wang J, Liu L, Ma J, Shen B (2020) Preparation of additive-free glass-ceramics from MSW incineration bottom ash and coal fly ash. Constr Build Mater 254:119345

Zimmer A, Bergmann CP (2007) Fly ash of mineral coal as ceramic tiles raw material. Waste Manag 27:59–68

Chapter 4
Environmental Effects of Coal and Biomass Ash Generation

Abstract There are apprehensions of presence of potentially toxic organic and inorganic substances in coal and biomass ash. Potentially toxic elements, PAHs, PCBs, and radioactive elements are some of the environmental concerns associated with the handling, utilization, and disposal of these ashes. During combustion in the power plant, volatile toxic elements are released from the coal and significant part of the element gets adsorbed on the ash particles at cooler zones of the furnace. These surface-associated metals are easily mobile and may get dispersed into the environment. Inorganic elements in the ash could be associated with the water-soluble phase, glass phase, bound to metal oxides or silicate matrix. The release of the elements is measured by leaching test. The concentration of elements in the leachate is compared with the values provided by regulatory authorities to evaluate the environmental acceptability of the ash. Weathering of ash or its interaction with water and CO_2 and other reactive liquids in the disposal or utilization sites leads to decomposition of glassy phase and formation of new clay phases. These new clay phases fix the potentially toxic elements in immobile pools. Thus, the pond ash or weathered ash has fewer risks than the fresh ash. Depending on the presence of unburnt carbon, halides, and the prevailing combustion conditions in the furnace, some ashes have considerable amount of PAHs, PCBs, and different dioxins. But studies with these organic pollutants are relatively less. Construction materials prepared from ash could exhibit radiation hazard, however, most of the studies showed radiation levels within the permissible limits. All these potential hazards from ash depend on the nature of the ash, so judicious screening and adoption of appropriate abatement techniques has to be followed while utilizing the ashes for different gainful purposes.

4.1 Toxic Components in Ash

4.1.1 Organic Pollutants

As detailed earlier, coal consists of organic and inorganic components. The organic component of coal consists of two sub-components: (a) the macromolecular, insoluble (by normal organic solvent) dominant sub-component, comprising a framework

A. K. Singh et al., *Ash from Coal and Biomass Combustion*,
https://doi.org/10.1007/978-3-030-56981-5_4

of condensed aromatic and hydro-aromatic elements; (b) less dominant portion of organic matter that are soluble in normal organic solvents. It is this soluble component that is marked by varying proportions of aliphatics, polycyclic-aromatics and hydro-aromatic-hydrocarbons, hydroxylated-polycyclic-aromatic compounds, and hetero-cyclic compounds. With increasing rank, the aromaticity of the organic structure in coals also increases. During the process of coal combustion, both the soluble and non-soluble components of organic matter in coals come into play, and depending upon the rank the combustion performance is affected. The poly-aromatic hydro-carbon (PAH) compounds present within coal are of environmental and human health concerns, and may concentrate in the ash fractions or the gaseous fractions produced from coal combustion (Verma et al. 2015).

PAH is a broad phrase for organic compounds comprising two and/or more allied benzene rings, and are known as toxic atmospheric pollutants as they comprise members which are known to be mutagenic or carcinogenic (Liu et al. 2000). PAH in coal becomes readily available during cooking, pyrolysis, and combustion processes. Incomplete combustion of organic matters in coal or other fossil fuels has been believed to be the most important factor that results in formation of PAHs (Costa and Sauer 2005; Müller et al. 1977). Consequently, earlier works have shown that PAH concentration in coal fly and bottom ash correlates with combustion conditions and the character of the coal being used in the power plant (Liu et al. 2000). While PAH from coal combustion residues has been long known, unburnt coal as a source of PAH leading to contamination of associated sediments/soils was not thought upon till recently. Coals present in coal seams are also known to have PAH concentrations in the range of hundreds of mg/kg, and occasionally up to thousands of mg/kg coal (Stout and Emsbo-Mattingly 2008). Further, the PAHs present in raw or feed-coal are known to affect the performance of the catalysts used in the refinery (Chen et al. 2005). The United States Environmental Protection Agency (EPA) has identified and prioritized 16 PAHs as perilous air pollutants, ranging from 2 to 6 condensed aromatic rings (Fig. 4.1).

Two processes, viz., pyrolysis and pyrosynthesis, have been opined by several authors that they control PAH formation and release during the combustion process (Chen et al. 2005; Mastral et al. 1999; Williams et al. 1986). Smaller, unstable, and highly reactive organic fragments produced due to breakdown of organic compounds during pyrolysis produce stable PAH by pyrosynthesis. The characteristic properties of the bulk coal samples also play a strong role in the amount of PAH species produced. Earlier works have documented that abundance of PAH in coals correlates positively with its rank (Xue et al. 2006). Similarly, organic sulfur compounds have also been observed to be higher in igneous-intrusion affected coal seams. However, the relationship between PAH concentration and coal rank is not straightforward. Earlier studies have revealed that around second coalification jump, that is, at vitrinite reflectance values of 0.9 and 1.1%, maximum aromatic-hydrocarbon yield and EPA-PAH are observed, respectively (Radke et al. 1990; Zhao et al. 2000). However, at higher ranks, beyond vitrinite reflectance values of 1.1%, the concentration of 2–6-ring PAHs is reduced, and an increase in the concentration of 6 or higher-ringed compounds takes place (Chen et al. 2005). Similarly, with changing coalification,

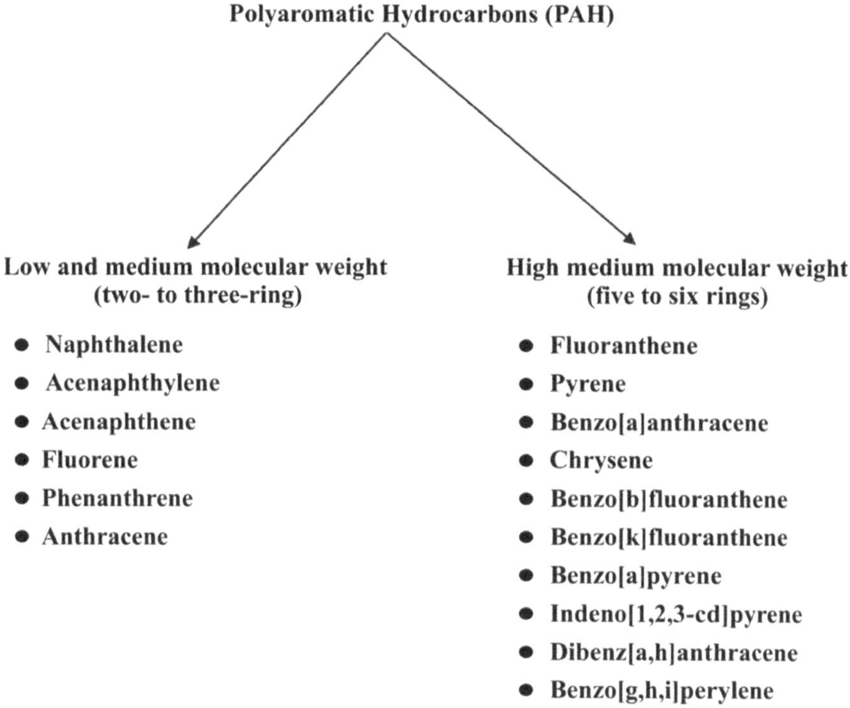

Fig. 4.1 United States Environmental Protection Agency (EPA) identified/prioritized 16 PAHs

PAH patterns also undergo certain variations. At the high-volatile bituminous stage of coal maturity, bicyclic and tricyclic PAHs are commonly present and show structural link with the parental molecules present in the coal (Puettmann and Schaefer 1990). As already mentioned at higher ranks, greater number of ring structures prevails, and this is also coupled with the lowering of oxygen content at higher ranks which results in removal of hydroxylated-aromatic compounds (viz., phenol), which are more common at lower ranks. At higher ranks, the PAH patterns become less complex (Puettmann and Schaefer 1990).

Earlier work of Verma et al. (2015) on coal, fly ash, and bottom ash samples from Unchahar Thermal Power plant, India showed that total PAH content in feed-coal to be much higher than the concentrations in the fly ash and bottom ash fractions. They also importantly inferred that higher molecular weight (4–6 rings) PAHs are more abundant in the coal and in the produced fly ash component, while the bottom ash fraction was marked by the presence of low molecular weight compounds (2–3 rings). For the coal and the fly ash samples, the higher concentration was shown by high molecular weight PAHs followed by medium molecular weight PAHs, and the least being low molecular weight PAHs. On the contrary, for the bottom ash samples, the trend was observed to be exactly reverse. The PAHs contents in the ash are much lower than the respective content in the feed-coal (Table 4.1).

Table 4.1 Distribution of PAHs in the feed-coal, fly ash, and bottom ash of a coal fired power plant

PAH compounds	PAH content (mg/kg)		
	Bed ash	Fly ash	Feed-coal
Naphthalene	0.035	0.111	0.059
Acenaphthylene	0.024	0.017	0.010
Acenaphthene	0.009	0.007	0.016
Fluorene	0.004	0.010	0.028
Phenanthrene	0.029	0.018	0.082
Anthracene	0.023	0.015	0.080
Fluoranthene	0.044	0.027	0.953
Pyrene	0.042	0.016	0.930
Benzo[a]anthracene	0.007	0.012	1.420
Chrysene	0.001	0.011	1.404
Benzo[b]fluoranthene	0.011	0.017	5.758
Benzo[k]fluoranthene	0.009	0.012	3.224
Benzo[a]pyrene	0.026	0.027	4.599
Dibenzo[a,h]anthracene	0.005	0.004	7.341
Indeno[1,2,3-c,d]pyrene	0.000	0.001	2.040
Benzo[g,h,i]perylene	0.000	0.003	7.325
Total PAHs	0.270	0.307	35.271

Recent years have also seen the usage of municipal solid wastes as a potential feedstock to generate energy through incineration. Previous works of (Dyke et al. 2003; Junk and Ford 1980; You 2008) have revealed that a range of PAHs are formed due to incineration of municipal wastes. You (2008) observed in their study that PAH from coal combustion was less than when coal and municipal solid waste (MSW) were used together. With increasing content of MSW in coal–MSW mixtures, low-ring PAHs were observed to become dominant in the emissions.

Similarly, biomass ashes produced from biomass combustion are also marked by the presence of PAHs (Masto et al. 2015), and its formation is related to the biomass-type (physical and chemical nature) and the environment of combustion (Ross et al. 2002). In a work on wood ash, Sarenbo (2009) discussed the presence of PAHs in wood ash of variable concentrations. Košnář et al. (2016) observed higher PAHs in fly ash compared to bottom ash from biomass combustion. Similar observations were also made by Masto et al. (2015), on their work on ash from different types of biomass collected from power plants. In contrast to the pattern of PAHs in coal ash, the biomass fly ash showed the presence of lower molecular weight PAHs, while the bottom ash showed the presence of higher molecular weight PAHs.

Like PAHs, polychlorinated biphenyls (PCBs) are persistent pollutants present in the ash. The tetra chloro-substituted CB77 congener was the abundant PCB present in fly ash (Sahu et al. 2009). In another study it was found that low molecular weight PCBs are abundant in the ash with an average value of the total of 12 dioxin-like PCBs was 2.57 ng/g (Li et al. 2017). Municipal solid waste combustion ash contains significant amount of polychlorinated dibenzo-p-dioxins, polychlorinated dibenzofurans, and dioxin-like polychlorinated biphenyls (Pan et al. 2013). Dioxins and PAHs are widely studied for incineration ashes (Fan et al. 2019; Fujimori et al. 2011; Mubeen et al. 2017; Till et al. 1997); data on coal and biomass ashes are few. The content of these compounds is relatively lower in the ash obtained from gasification of straw (Asikainen et al. 2002).

4.1.2 Inorganic Pollutants

It is their content of toxic elements and heavy elements of fly ash, in addition to PAH, which has been a point of worry for technologists and policy makers. Similar to the generation of PAHs, during coal combustion, the toxic elements may be concentrated in the fly ash and bottom ash components. A comparison of the content of potentially toxic element between coal, lignite, and biomass ash shows that the plant nutrient elements like Cu and Zn are higher in biomass ash; and typical enrichment of Cd in biomass ash (Table 4.2). A range of toxic elements, viz., aluminum, lead, cadmium, chromium, mercury, vanadium, boron, arsenic, selenium, cobalt, nickel, uranium, thorium, and so on have been reported to be present within coal and fly ash and are considered as toxic elements that can impact environment and health of those being exposed to it (Finkelman et al. 2002; Jambhulkar et al. 2018; Verma et al. 2015). Earlier works have shown that certain compounds of arsenic, cadmium, chromium, and nickel are known to be carcinogenic for humans (Meij 1994). It is for these reasons, the fly ash elemental composition and probable toxicity have been extensively studied by several researchers. The fly ash toxic composition is largely dependent on the initial coal composition (which is dependent on its depositional environment) and the conditions effective during combustion, volatilization,

Table 4.2 Content of potentially toxic elements in coal, lignite, and biomass ash [Adapted from (Masto et al. 2015; Ram et al. 2014; Singh et al. 2011)]

Elements	Content of potentially toxic elements (mg/kg)		
	Coal ash	Lignite ash	Biomass ash
As	3.46	0.19	1.80
Cd	1.40	1.28	18.00
Co	3.88	19.81	5.2
Cr	47.10	44.15	51.00
Cu	45.80	119.80	241.00
Ni	11.70	28.31	64.00
Pb	44.10	55.07	2.00
Zn	53.00	19.85	400

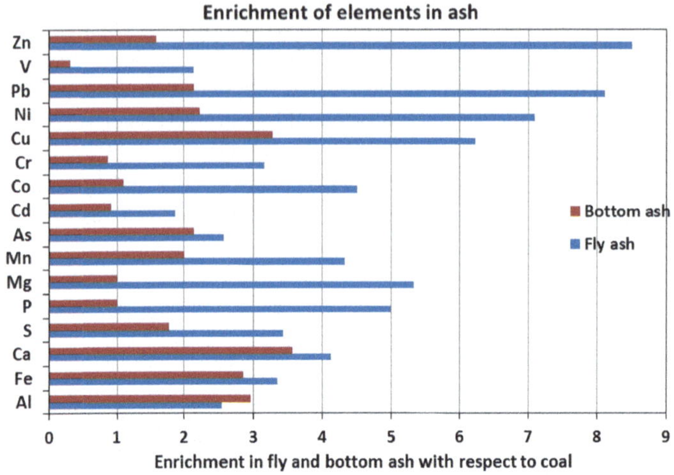

Fig. 4.2 Enrichment of major and trace elements in fly ash and bottom ash

and condensation process during operation, and particle size distribution of the ash (Meij 1994; Ram et al. 2014; Singh et al. 2011).

In their study Verma et al. (2015) observed arsenic, cadmium, cobalt, chromium, copper, nickel, lead, vanadium, and zinc to be more enriched in fly ash compared to the bottom ash component and the coal sample itself (Fig. 4.2). In general, the fly ash component was observed to be rich in several trace elements, while some (viz., manganese, cobalt, nickel, etc.) were observed to be in greater concentrations within the bottom ash fraction. Finer particle size and surface-to-volume ratio of fly ashes have been attributed to be the reason why they are observed to be typically rich in different elements (Raclavska et al. 2009). In their work, they further observed that concentrations of certain elements such as arsenic, nickel, copper, lead, and zinc were significantly higher. Earlier studies have opined that during combustion, these elements typically undergo volatilization and later undergo condensation on the surface of fly ashes, thereby enriching the fly ashes instead of the bottom ashes (Steenari et al. 1999; Yan et al. 2001). Arsenic often undergoes vaporization at low-temperatures, and partitions into the fly ash fraction, thereby increasing the toxicity of fly ashes (Dahl et al. 2009). Previous works pertaining to the behavior of trace elements and their mobility 'reactions' caused during coal combustion have revealed that volatility is controlled by the reactivity of the elements with sulfur or other elements with volatility, combustion time-temperature, and the kind of ash generated (Meij 1994). One of the most critical concerns about the toxic elements in fly ash is that when the fly ashes come in contact with water bodies, they may be leached out and be released into the atmosphere (Jambhulkar et al. 2018; Wang et al. 2020). Presence of arsenic in fly ash has drawn significant attention among researchers, mainly due to the toxic effects of arsenic. It is believed that arsenic content is related to pyrite contents in coal, which are decomposed during coal combustion and concentrated in the fly ash. Arsenic and other oxyanions are generally more leachable in alkaline-medium and can cause significant worries when they enter the human chain system through soil or crops. Among the other elements of 'worry', radionuclides are also known to be present in coal ashes, although in low concentrations (Papastefanou 2008). Although uranium and thorium have been reported to be present in coal fly ash, they are generally in extremely low concentrations, but may have deleterious impact on the surrounding environment. For example, when coal fly ash containing the naturally occurring radionuclide (NOR) is blended with building materials for construction, it could result in increasing the radioactivity levels of building materials (Roper et al. 2013). Thorough knowledge on this aspect is thus extremely important for human health safety. However, leachability of NORs before being used for construction could cause severe environmental problems (Chakraborty and Mukherjee 2009). Appropriate methodologies to abate the leachability of toxic components from fly ash, especially from dump sites is thus an extremely vital and critical step. Adding alkaline materials, viz., $Ca(OH)_2$, CaO are known to deter the toxic component leaching from fly ash (Wang et al. 2020). Other toxic elements and heavy metals, viz., manganese, boron, arsenic, molybdenum, mercury, chromium, cadmium, cobalt etc. can enter the food-chain when such fly ashes are used for soil amendment, and can thus have deleterious impact on human beings and the associated ecosystem (Jambhulkar et al. 2018). Certain works have revealed that fly ash can have genetoxicity on plants and can thus impact plants adversely when used for soil amendments (Chakraborty et al. 2008; Jana et al. 2017).

Radionuclide content was measured for wide varieties of fly ash generated from different biomasses. The activity varied from 50 to 2000 Bq/kg. The most abundant nuclide was ^{137}Cs, followed by ^{40}K, ^{90}Sr, ^{210}Pb, ^{226}Ra, ^{232}Th, ^{134}Cs, ^{235}U (Rantavaara and Moring 2000). The radioactivity level in vast number of coal ash is not significant when compared to soil/rock references. So no health concerns from radionuclide in ash (Zielinski and Finkelman 1997). Fly ash and constructions made by fly ash was evaluated for radiations and the existing data showed no significant increase in gamma radiation. Radioactive elements are generally held within the glass portion of fly ash (Temuujin et al. 2019)

4.1.3 Human Health Hazards

Fly ash can affect the health of human beings on consumption of contaminated plant produces grown on fly ash added soils; leaching of potentially toxic substances to groundwater used for drinking; inhalation of super fine ash particles and by radiation emitted by fly ash-based building materials.

Leaching studies with different fly ash showed that the concentration of Ba, Cr, Mo, Se, and SO_4 exceeded the permissible limits and affected the bioluminescence of the test organism, *Vibrio fischeri* (Tsiridis et al. 2012). Growth and germination of all the tested plant species (barley, oats, rye, wheat, regreen, and triticale) was affected by fly ash or bottom ash growth media. Concentration of all elements was higher in the fly ash and bottom ash added media (Bilski et al. 2012).

Experiments with rat showed that inhalation of ash particles get translocated to extra-pulmonary organs. Ash deposit has shown toxic effects on different tissues (Mani et al. 2007). Humans living near the power plants are exposed to fugitive dusts and ash-contaminated drinking water. The exposure route may be ingestion, dermal contact, and inhalation. Leaching of potentially toxic elements from ash and its subsequent contamination to soils and water affects the health of the community (Flues et al. 2013; Jones et al. 2012; Patra et al. 2012b; USEPA 2010). Children are more exposed to ash due to the increased respiratory activity, more hand to mouth activity, and playing in grounds near power plant and disposal sites. Fine and ultra-fine particles of the ash can easily pass through the nasal pathway and enter into circulation system and brain (Block and Calderón-Garcidueñas 2009; Calderón-Garcidueñas et al. 2007). Continuous exposure to particles increases the cytokines content in the body. Many metals in the ash are neurotoxic and create neurological dysfunctions in the early stage of development (Rodríguez-Barranco et al. 2013). Living near ash storage sites causes sleep disturbances to children due to increased exposure to potentially toxic elements and ash particles (Sears and Zierold 2017). Respiratory diseases were higher for adults living near the ash storage facility (Hagemeyer et al. 2019). Occupational exposure to workers handling As-rich ash increased malignancy-caused death than the workers handling low As ash (Bencko et al. 1980, 1988). Other studies found occupation exposure-related DNA damage (Chen et al. 2010) and higher plasma malondialdehyde (Liu et al. 2008) for worker handling fly ash than bottom ash. People living near thermal power plants are more likely to suffer from respiratory problems, gingiva, and skin symptoms (Zierold et al. 2020). Epidemiological exposure studies on fly ash are limited (Kristina and Chisom 2020).

Exposure risk from the potentially toxic elements in soils near a coal-fired power plant is depicted in Table 4.3. For all the elements, the concentrations in the surrounding soils are within the permissible limits.

Table 4.3 Exposure risks to potentially toxic elements present in soils near a coal-fired power plant

Element	Content in soil (mg/kg)	Exposure risk (HQ)				
		Ingestion	Dermal contact	Inhalation	Soil-plant	HI = ΣHQi
As	3.23	1.0E-02	1.6E-04	3.5E-07	4.3E-03	1.5E-02
Co	12.8	4.1E-02	6.4E-04	3.4E-06	1.7E-02	5.9E-02
Cr	147	4.8E-02	3.0E-02	2.4E-06	2.0E-02	9.7E-02
Cu	25.5	6.1E-04	9.6E-06	–	2.6E-04	8.8E-04
Mn	478	3.3E-03	5.2E-05	1.6E-05	1.4E-03	4.7E-03
Ni	58.2	2.8E-03	1.1E-03	1.1E-06	1.2E-03	5.1E-03
Pb	20.2	5.6E-03	8.7E-05	–	2.3E-03	8.0E-03
Zn	71.8	2.3E-04	3.6E-06	–	9.7E-05	3.3E-04

4.2 Leaching Studies

The utilization of fly ash in different sectors is detailed in the previous chapters. Fly ash also contains some potentially toxic elements that may get released into the environment and affects animal, plant, and human health. Some previous studies reported hazards from leaching of As from the ash (Wang et al. 2020). There are several leaching protocols to evaluate the ash. Leachate is the liquid portion that percolates through the ash pores and contains dissolved solids and suspended solids. Leaching test may be static, single addition of test solution; or dynamic, multiple addition of reacting solution. There are batches as well as continuous leaching tests. But till date there is no single universal test method to be used for all kinds of solid waste and environmental settings. Selection of appropriate test methods will provide useful information for taking decision on ash disposal and utilization. According to Tiwari et al. (2015) leaching methods are designed for three purposes (Table 4.4). The laboratory test may not represent the real release of chemicals from ash in the environmental conditions, because these are mostly used for regulatory purposes (Kosson et al. 2002).

In thermal power plants, fly ash is generally disposed in ash pond in the form of slurry. The leaching potential of the ash will give important information on the potential percolation to groundwater and transport to the nearby environment (Praharaj et al. 2002). The release of soluble components from ash depends on the pH of the ash and many other parameters like coal properties, chemical environment of the disposal site, elemental and mineralogy composition of the ash, physical properties, dissolved phases in ash, combustion conditions in the boiler, volatilization, cooling, melting, and neo formation of new mineral and chemical species in the boiler (Jankowski et al.

Table 4.4 Different purposes of leaching test

	Leaching test	Objective
1	Simulate contaminant release	Laboratory tests conducted to simulate the liberation of elements under particular environments. For example, USEPA's TCLP method is a kind of synthetic acid rain for leaching the potentially toxic elements and other chemical constituents from ash/solid waste
2	Understanding fundamental leaching pattern	To get fundamental data on the concentration and chemistries of different chemical constituents in the leachate and the solid residue
3	Sequential chemical extraction	This protocol is followed to understand the chemical association of elements in the ash. Generally, the elements are grouped into water-soluble, exchangeable, carbonate bound, metal oxide bound, organic bound, and residual fractions

Table 4.5 Description of different leaching procedures

	Leaching method	Description
1	Column leaching test	Replicating the flow of percolating water through a porous bed. The assumption is the leaching solution is uniformly distributed in the column and all ash particles uniformly interact with the leaching solution. The flow direction may be upward or downward. This test can be performed under saturated and unsaturated settings
2	Batch leaching tests	Sample is allowed to interact with the known leached volume for a particular time phase. Mostly the system is agitated for better interaction. After the agitation/reaction time, the filtrate is separated and analyzed. The liquid-to-solid ratio (L/S), pH of the reacting solution, time of reaction are some of the variables that could be modulated to suit the objective of the test
3	Standard leaching tests	Commonly available standard test protocols are: 1. Toxicity characteristic leaching procedure (TCLP) 2. Extraction procedure toxicity test (EPTOX) 3. Synthetic precipitation leaching procedure (SPLP) 4. Standard test method for shake extraction of solid waste with water (ASTM-D3987) 5. California waste extraction test (WET) 6. Gastric juice simulation test (GJST)

2006; Kim 2002). Single-step leaching test could not give temporal variation in the release of elements, but column leaching test can give information on the temporal behavior. Table 4.5 indicates the different leaching test available, and depending on the nature of the requirement appropriate test methods are to be adopted.

According to Tiwari et al. (2015), the choice of the leaching analysis depends on physical condition of the ash; environment in which it has to be disposed; requirement of the local regulatory authority; and the chemicals/pH used in the test protocols. pH is the most important leaching condition; and acidic pH releases more elements.

In the fly ash leachate, the concentration of Ca, Na, K, Mn, Fe, S, and Pb was relatively higher, while Cd, Mg, Cu, Cr, Zn, and Co was minimal. Low leaching of elements is due to the alkaline pH of the ash (Ugurlu 2004). Leaching of As and Se was higher for bituminous coal ash, compared to the ash from sub-bituminous coals. Compared to arsenic, selenium was more readily leachable for these two ashes. Calcium precipitation decreased the leaching of As and Se (Wang et al. 2009). For biomass ash, K and P are the major composition of the leachate; content of all the trace elements (except Cr) was within the regulatory limits (Maresca and Astrup 2014). There are several works on the leaching of potentially toxic elements from coal and biomass ash. The quantity of elements leached depends on the ash properties, L/S ratio, pH of the extracting solution, and the chemical composition of the interacting liquid.

The concentration of the potentially toxic elements is generally compared with the permissible limits provided by regulatory authorities. An example of such evaluation is given in Fig. 4.3.

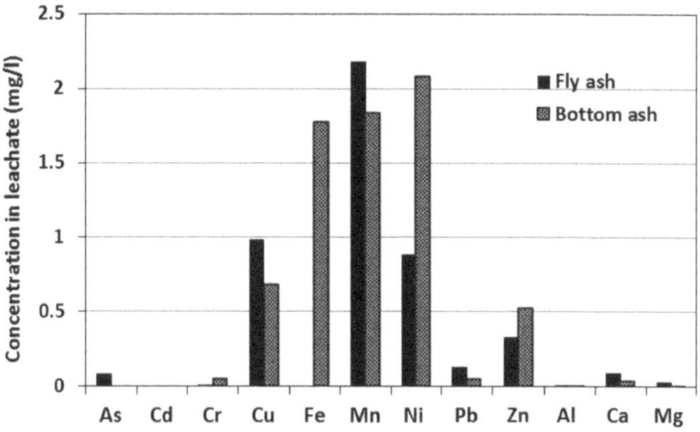

Permissible limit (mg/l) [Drinking water quality Indian Standards IS 10500]

As	Cd	Cr	Cu	Fe	Mn	Ni	Pb	Zn	Al	Ca	Mg
0.05	0.003	0.05	1.5	0.3	0.3	0.02	0.05	15	0.2	200	100

Fig. 4.3 Concentration of elements in the leachate from leaching of elements from fly and bottom ash

4.3 Chemical Speciation and Weathering

Coal and its ash are heterogeneous. During combustion in the boiler, the elements present in the fuel feed (coal or biomass) undergo different transformations and these reactions continue even in the post combustion zones. The different processes are decomposition, volatilization, fusion, agglomeration, and condensation (Jones 1995). Volatile elements (As, Hg, Cl, Cr, Se, B, and S) get condensed on the ash surface and mostly they are soluble and associated with Ca. These condensations make the relative enrichment of these elements into finer particles and there are solubility gradients across the ash particles. Along with these volatile elements, Cd, Cu, Mo, Sb, V, and Zn are also enriched on ash surfaces; these surface-enriched elements are prone to leaching (Iyer 2002). Some of the elements like Ba, Co, Cr, Mn, Ni, and Pb are enriched in the core and these elements are not easily leached (Kukier et al. 2003). In general, Ca, SO_4, Cl, Na, and K are soluble across pH and are prominently leached from ash. pH of the ash-solution mixture is the main contributing factor for the leaching of element. The pH depends on the content and balance between Ca and S in ash. Ca-rich ashes are alkaline and maintain a very high pH with low solubility of potentially toxic elements, whereas S-rich ash produces sulfuric acid in the presence of water and dissolves from elements present in ash. The pH of the ash–water system is dynamic and the solubilized element may re-precipitate depending on the changes in pH (Izquierdo and Querol 2012). The leaching potential of the major elements and their chemical association in ash are depicted in Table 4.6.

In coal ash most of the potentially toxic elements (Co, Cr, Cu, Pb, Ni, and Zn) are associated with the insoluble silicate matrix (Fig. 4.4), whereas Ca, Mg, K, and Na

Table 4.6 Elemental association in coal and biomass ash [Adapted from (Izquierdo and Querol 2012; Höfer et al. 2017; Masto et al. 2015; Ullah et al. 2017)]

	Element	Association		Solubility
		Coal ash	Biomass ash	
1	Al	Insoluble and associated with the glassy phase	Associated with potassium and sodium alumino silicates	Maximum solubility at pH <4; and 9–11; some amorphous phase with open structure get solubilized
2	Ca	Associated with lime/anhydrite/calcite and within the glassy phase	Oxides, sulfates, carbonates, and silicates of Ca. Also exist as complex compounds with K, P, and Mg	Most of the Ca is soluble and easily leachable; some fraction is sparingly soluble
3	Fe	Magnetite mixed with hematite; minor portion in glassy phase and as FeOH	Amorphous phase	Soluble only at very low pH. Leaching of Fe is generally less
4	Mg	Glassy phase	MgO and Ca–Mg silicates	Soluble only in acidic pH
5	P	Ca phosphates and in silicate matrix	Ca and Ca-K phosphates	Extremely insoluble
6	K	Feldspar, clay, and tightly bound in glassy matrix	Potassium sulfate, carbonate, Ca/K salts and K-Al silicates	Some soluble compounds are leachable. More soluble fraction in biomass ash than coal ash
7	Na	Glassy matrix, feldspar, clay	Na–Al silicates	Some components are soluble. More soluble fraction in biomass ash than coal ash
8	Si	Glass, quartz; weathering enriches Si in the glass phase	SiO_2, silicates of Ca, K, Mg, Na	Soluble in highly alkaline and acidic pHs
9	S	Surface association	Sulfates of Ca, K, Mg, Sa and their mixed phases	Most of the S compounds are soluble
10	Ti	Retained in aluminosilicate glassy matrix	–	Not soluble
11	As	Surface of ash particles	Bound to Fe/Mn oxides	For acidic fly ash, As solubility increases with increase in pH; whereas it is reversed for alkaline ash
12	B	Dominant surface association	Carbonates and metal oxide	Highly soluble

(continued)

Table 4.6 (continued)

	Element	Association		Solubility
		Coal ash	Biomass ash	
13	Cd	Surface of ash particles	Aluminum Cd silicates	Soluble in acidic pH
14	Cu	Glassy matrix, Cu–Fe sulfides	Organic fractions, sulfides, chalcopyrite	Soluble at low pH and sensitive to oxidation reduction reactions
16	Cr	Alumino silicates.	Fe/Mn oxides.	Soluble at very low pH and at reducing condition.
17	Co	Fe bearing species	Organics and metal oxides	Poor solubility near neutral pHs
18	Mn	Ferromagnetic and glassy phase	Alumino silicate phases	Insoluble near neutral pHs
19	Ni	Magnetic and silicate phase	Silicates and metal oxide	Soluble at low pHs
20	Pb	Surface and internal glassy and silicate phases	Sulfates and oxides of Pb; as well as in alumina silicate phases	Not easily soluble
21	V	Surface association, glass and magnetic	–	Soluble in alkaline condition
22	Zn	Mostly associated with glass	Zn oxide and silicates	Amphoteric
23	Se	Completely volatilized and may form stable compounds with Ca	Ca–Se compounds	Soluble in acidic and alkaline pHs
24	Hg	$HgCl_2$ on ash surfaces	Ash surfaces	Very low concentration in the leachate

are distributed in water-soluble, exchangeable fractions with increasing proportions for lignite and biomass ash. In general, most of the elements of coal and lignite ash are present in recalcitrant pools that are not easily leached in the environment. In case of biomass ash some elements like Cu, Sr, K, Ba, and Ni are present in relatively soluble fractions.

Fly ash is used in different sectors, despite that huge amount of ash is unutilized and stored in ash ponds. Dry handling and disposal of ash pose harmful exposure risks to the community. The release of heavy metals from the disposed fly ash after weathering is one of the major challenges if the metals are released which may contaminate the groundwater and the nearby environment. There are reports on the high level of toxic metals that could affect the human health through inhalation of the ash particles or through drinking the ash leachate contaminated water (Akinyemi et al. 2019). In natural setting the disposed ash reacts with water and atmospheric CO_2 to form different mineral phases. Formation of new clay minerals (Bradley et al. 1999) like illite from the original glassy phases has been reported. However,

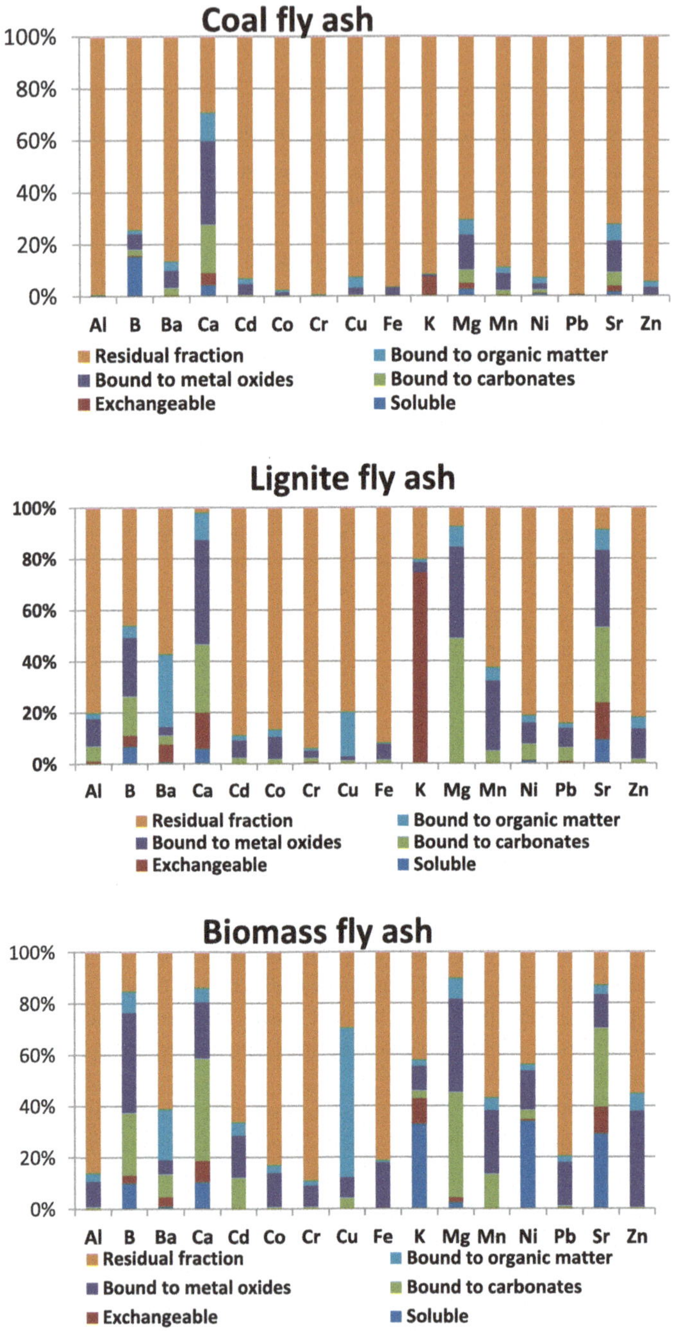

Fig. 4.4 Distribution of elements in different chemical pools

the time period required for this mineralogical transformation is about 10 years (Zevenbergen et al. 1994). Upon reaction with water, initially the soluble salts of the ash is released followed by dissolution of glassy matrix (Warren and Dudas 1984). Long-term weathering studies show that the metal concentration is enriched in water-soluble pools for fresh ash, but for a weathered ash the metal in water-soluble pool is significantly low. In weathered fly ash, most of the trace elements are enriched in the non-mobile silicate phases. In weathered fly ash there is enrichment of new glass, iron oxides, calcite, and amorphous phases and they are more reactive than the fresh ash. After weathering, some of the soluble phases of Na, K, and Ca are flushed out from the ash. Overall, most of the potentially toxic elements are less mobile in the weathered ash and they pose less risk (Akinyemi et al. 2020). Eco toxicological studies showed that the weathered fly ash are non-toxic and can be utilized for different applications (Bandarra et al. 2020; Bradley et al. 1999). Rate of weathering of coal ash is faster than volcanic ash; upon weathering formation of clay minerals fixes the potentially toxic elements in the non-mobile phases (Zevenbergen et al. 1999).

4.4 Toxic Substance in Products Prepared from Ash

Leaching of potentially toxic elements from concrete prepared using fly ash was studied by Gwenzi and Mupatsi (2016). The concentration of Fe, Zn, Cu, Mn, and Pb in the leachate was lower and comparable with the conventional concrete. So there is lesser harmful effect from fly ash used concrete. Column and batch leaching test of fly ash concrete showed that the concentration of all the elements in the leachate are within the allowable caps (Rankers and Hohberg 1991). Leaching of heavy metals from fly ash bricks was less than that of clay bricks (Gupta et al. 2019). Lysimeter studies on the leaching of heavy metals from fly ash road revealed the heterogeneity of the leachate data; and except for few sites the leachate concentration was comparable with the no fly ash control (Lindroos et al. 2019). Fly ash carbonated concrete has showed that carbonation is not always beneficial in restricting the leaching of Cu and Pb (Zha et al. 2019). In fly ash stabilized soil, the leaching of metals was controlled by Cao and MgO present in ash (Leelarungroj et al. 2018). Leaching of Fe, Ni, and Mn from lightweight fly ash bricks were higher than the drinking water standards (Cengizler 2009). Leaching from fly ash brick is negligible (Liu et al. 2009) and within the limit (Sarode et al. 2010).

There is a general apprehension that building made of fly ash emits more radiation hazard. The natural radioactivity in fly ash is higher, however, the building constructed with ash brick has radon concentration well within the action level of 200–600 Bq/m^3. Fly ash use in construction of buildings did not show any radiological hazard (Santhanam et al. 2005). Another study from Greece also observed that the natural radioactivity (^{226}Ra, ^{232}Th, and ^{40}K) was higher for fly ash than other building material, however, the activity concentration index (I) was within the permissible limit γ-radiation exposure (Papaefthymiou and Gouseti 2008). Glazed tile, roof tile, and other cement products prepared from fly ash have the radium equivalent activity and external hazard index higher than the permissible limit (Lu et al. 2014). Similar high radiation level was reported by Lu and Zhang (2007). Radon exposure calculated for fly ash brick building exceeded the acceptable limit (Chauhan et al. 2003). Radon concentration inside the fly ash building near power plant site is comparable to the clay brick building and did not exceed the action limit, except for one site (Bala Sundar et al. 2008). The literature on the radiation hazard has shown both positive (Abdullahi et al. 2019; Hatungimana et al. 2020; Karim et al. 2019; Sanjuán et al. 2020) and negative (Mehra et al. 2019) results, so proper monitoring of potential radiation exposure from fly ash building is essential.

Concentration of heavy metals in *Beta vulgaris* increased with increase in concentration of fly ash dose. The yield also decreased with ash dose (Singh et al. 2008). However, Rautaray et al. (2003) observed decrease in metal content in rice grain with increase in fly ash dose. This is due to the increase in soil pH with ash application. Heavy metals present in fly ash caused oxidative stress to chick pea plant (Pandey et al. 2010). High dose of fly ash application increased toxic levels of heavy metals in rice grain (Singh et al. 2016). Patra et al. (2012a) reported that rice grain yield was increased up to 40% by application of 200 t/ha ash; but it increased As content in the grain by 1.8 times. Based on a detailed analysis of the published data, it has been observed that fly ash application up to 5% maintained reasonable level of plant nutrients without any accumulation of potentially toxic elements in the plant. At higher rates of ash application, elements especially B and Se accumulation increased (Yu et al. 2019). Use of weathered ash (pond ash) increases the crop yield and the accumulations of potentially toxic elements were within the permissible limit (Fig. 4.5). So, for safer agricultural application, use of weathered pond ash should be recommended.

Contaminated wood ash increased the accumulation of heavy metals in rice, whereas normal wood ash did not increase the metal content in rice grain (Beesley

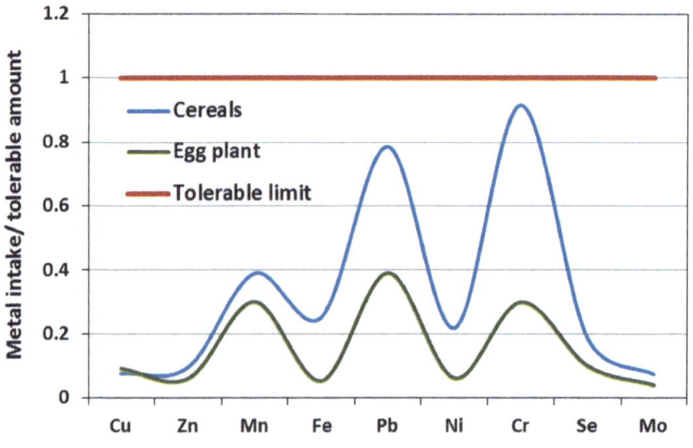

Fig. 4.5 Dietary intake of metals through crop produce grown with pond ash

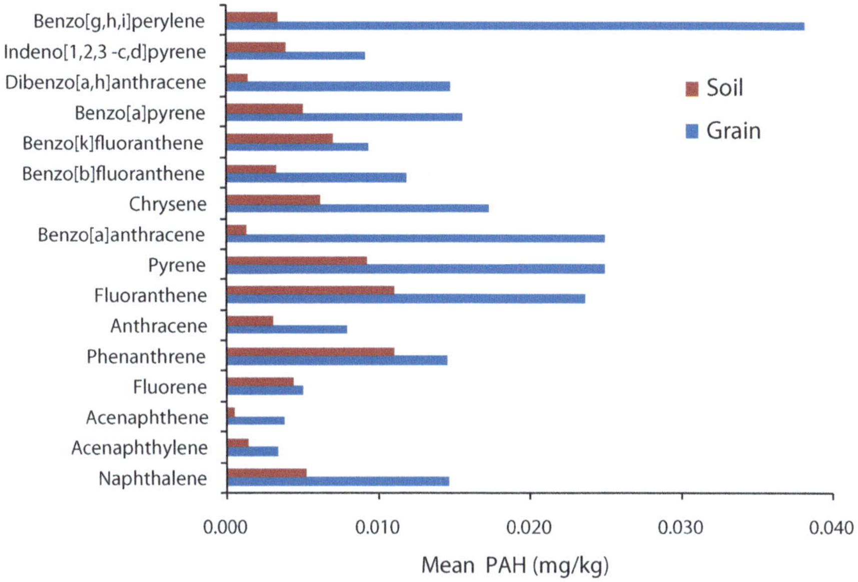

Fig. 4.6 PAHs content in the soil and crop produce growing near a coal-fired power plant

et al. 2019). Accumulation of heavy metal in crop produce due to wood ash application was reported (Meharg et al. 2009; Mollon et al. 2016; White and Broadley 2009).

Studies on the uptake and accumulation of PAHs in crop produce due to ash application are limited. The PAHs signals in soil and paddy grain near a thermal power plant is shown in Fig. 4.6. The predominant PAHs in soil is pyrene, fluoranthene, and phenanthrene, However, in grain benzo [g, h, i] perylene, benzo[a]anthracene, pyrene, fluoranthene, and chrysene are the dominant PAHs. When compared to the soil concentration, significant bioaccumulation of PAHs in grains was observed for almost all the PAH compounds.

4.5 Abatement of Ash Toxicity

Several processes like thermal treatment, solidification, stabilization, vitrification, classification by granulometric size particles, and aging or weathering are employed to decrease the leaching of potentially toxic elements from ash. Most of these processes are expensive, however, weathering of the ash for 6–12 weeks is considered as most feasible method to decrease the mobility of elements (Chimenos et al. 2000, 2003). Weathering decreased the mobility of elements (Pb and Zn) due to the formation of insoluble oxides and carbonates. However, leaching of Al increased due to metal oxidation (Chimenos et al. 2000). Carbonization of ash was followed to decrease the leaching. Carbonation decreased the leaching of Cu from municipal solid waste incineration ash; Mo and Sb was not affected; Cr leaching increased (Van Gerven et al. 2005).

Washing of fly ash prior to disposal or utilization has been attempted by few researchers. Different washing agents like sulfuric acid (Kashiwakura et al. 2010); nitric acid (Mulder 1996); diethylenetriamine penta-acetic acid (Ishaq et al. 2013); citric acid (Huang et al. 2011). Depending on the target element to be washed and its chemical association in the ash, specific washing strategies could be designed. Simple water washing is effective to remove significant amount of soluble salts from biomass ash. While making brick from biomass ash, to avoid salt efflorescence in the brick, a pre-treatment of water washing will be helpful.

Regarding agricultural application of fly ash, co-application of ash with organic manures or biochar decreases the metal uptake by the crop produce. The effect of biochar addition along with fly ash on the accumulation of heavy metals is depicted in Table 4.7. Co-application of fly ash and biochar showed significant decrease in the content of most of the potentially toxic elements in the maize grain.

Phytoremediation of ash-contaminated sites have been studied in detail. Some tolerant plant species naturally growing in the ash pond sites could be easily adopted for reclamation of ash-contaminated sites. The following criteria are to be followed while phytoremediation of fly ash sites (Pandey et al. 2009):

(i) Selection of appropriate adoptable plant species.
(ii) Plant should have high biomass production capacity and tolerate adverse soil and climatic conditions.
(iii) Should accumulate metal and must be tolerant to high salt and metal concentration.
(iv) Active secretion of chelating agent to increase the metal availability for root uptake.
(v) Should be easily cultivated/propagated.

Table 4.7 Effect of co-application of biochar and fly ash on the accumulation of potentially toxic elements in maize grain

	Maize grain (mg/kg)			
	Control	Biochar	Lignite fly ash	Biochar + fly ash
Cd	0.016	0.012	0.021	0.008
Co	0.049	0.051	0.043	0.046
Cr	0.596	0.640	0.856	0.852
Cu	35.6	111	70	109
Mn	6.68	8.56	5.85	5.30
Ni	0.760	0.764	0.705	1.137
Pb	0.409	0.486	2.262	0.340
V	1.09	0.94	0.77	0.97
Zn	20.3	24.9	21.9	23.0

(vi) Select appropriate agronomic practices of spacing, irrigation, fertilizer application, weeding, etc.
(vii) Consortium of plant species should be used.
(viii) Ease of harvest and handling of the metal accumulated plant parts.

Ram and Masto (2010) compared the level of potentially toxic elements in coal ash generated from different countries and found that the feed-coal has significant role on the concentration of potentially toxic element in ash. Thus, judicious selection of the ash for the intended purpose will help to contain the dispersion of toxic elements, if any.

References

Abdullahi S, Ismail AF, Samat S (2019) Radiological characterization of building materials used in Malaysia and assessment of external and internal doses. Nucl Sci Tech 30:46

Akinyemi SA, Gitari WM, Petrik LF, Nyakuma BB, Hower JC, Ward CR, Oliveira ML, Silva LF (2019) Environmental evaluation and nano-mineralogical study of fresh and unsaturated weathered coal fly ashes. Sci Total Environ 663:177–188

Akinyemi SA, Gitari WM, Thobakgale R, Petrik LF, Nyakuma BB, Hower JC, Ward CR, Oliveira MLS, Silva LFO (2020) Geochemical fractionation of hazardous elements in fresh and drilled weathered South African coal fly ashes. Environ Geochem Health

Asikainen AH, Kuusisto MP, Hiltunen MA, Ruuskanen J (2002) Occurrence and destruction of PAHs, PCBs, ClPhs, ClBzs, and PCDD/Fs in ash from gasification of straw. Environ Sci Technol 36:2193–2197

Bala Sundar S, Danalakshmi B, Santhanam R (2008) Radon measurements in fly ash buildings near major thermal power stations in India using SSNTD. Radiat Meas 43:S392–S394

Bandarra BS, Gomes LA, Pereira JL, Gonçalves FJM, Martins RC, Quina MJ (2020) Assessment of hazardous property HP 14 using ecotoxicological tests: a case study of weathered coal fly ash. Environ Sci Pollut Res 27:20972–20983

Beesley L, Hough R, Deacon CM, Norton GJ (2019) The impacts of applying metal(loid) enriched wood ash to soils on the growth and elemental accumulation of rice. Expo Health 11:311–324

Bencko V, Symon K, Stálnik L, Bátora J, Vanco E, Svandová E (1980) Rate of malignant tumor mortality among coal burning power plant workers occupationally exposed to arsenic. J Hyg Epidemiol Microbiol Immunol 24:278–284

Bencko V, Wagner V, Wagnerová M, Bátora J (1988) Immunological profiles in workers of a power plant burning coal rich in arsenic content. J Hyg Epidemiol Microbiol Immunol 32:137–146

Bilski J, Jacob D, McLean K, McLean E, Soumaila F, Lander M (2012) Agro-toxicological aspects of coal fly ash (FA) phytoremediation by cereal crops: effects on plant germination, growth and trace elements accumulation. Adv Biores 3:121–129

Block ML, Calderón-Garcidueñas L (2009) Air pollution: mechanisms of neuroinflammation and CNS disease. Trends Neurosci 32:506–516

Bradley CZJ, van Reeuwijk L, Shyam A (1999) Clay formation during weathering of alkaline coal fly ash

Calderón-Garcidueñas L, Franco-Lira M, Torres-Jardón R, Henriquez-Roldán C, Barragán-Mejía G, Valencia-Salazar G, González-Maciel A, Reynoso-Robles R, Villarreal-Calderón R, Reed W (2007) Pediatric respiratory and systemic effects of chronic air pollution exposure: nose, lung, heart, and brain pathology. Toxicol Pathol 35:154–162

Cengizler H (2009) Toxic elements leachability tests on light weight fly ash bricks. Asian J Chem 21:2950–2956

Chakraborty R, Mukherjee A (2009) Mutagenicity and genotoxicity of coal fly ash water leachate. Ecotoxicol Environ Saf 72:838–842
Chakraborty R, Mukherjee AK, Mukherjee A (2008) Evaluation of genotoxicity of coal fly ash in Allium cepa root cells by combining comet assay with the allium test. Environ Monit Assess 153:351
Chauhan RP, Kant K, Sharma SK, Chakarvarti SK (2003) Measurement of alpha radioactive air pollutants in fly ash brick dwellings. Radiat Meas 36:533–536
Chen Y, Sheng G, Bi X, Feng Y, Mai B, Fu J (2005) Emission factors for carbonaceous particles and polycyclic aromatic hydrocarbons from residential coal combustion in China. Environ Sci Technol 39:1861–1867
Chen HL, Chen IJ, Chia TP (2010) Occupational exposure and DNA strand breakage of workers in bottom ash recovery and fly ash treatment plants. J Hazard Mater 174:23–27
Chimenos JM, Fernández AI, Nadal R, Espiell F (2000) Short-term natural weathering of MSWI bottom ash. J Hazard Mater 79:287–299
Chimenos JM, Fernández AI, Miralles L, Segarra M, Espiell F (2003) Short-term natural weathering of MSWI bottom ash as a function of particle size. Waste Manag 23:887–895
Costa HJ, Sauer TC (2005) Forensic approaches and considerations in identifying PAH background. Environ Forensics 6:9–16
Dahl O, Nurmesniemi H, Pöykiö R, Watkins G (2009) Comparison of the characteristics of bottom ash and fly ash from a medium-size (32 MW) municipal district heating plant incinerating forest residues and peat in a fluidized-bed boiler. Fuel Process Technol 90:871–878
Dyke PH, Foan C, Fiedler H (2003) PCB and PAH releases from power stations and waste incineration processes in the UK. Chemosphere 50:469–480
Fan Y, Ren M, Zhang H, Geng N, Li Y, Zhang N, Zhao L, Gao Y, Chen J (2019) Levels and fingerprints of chlorinated aromatic hydrocarbons in fly ashes from the typical industrial thermal processes: implication for the co-formation mechanism. Chemosphere 224:298–305
Finkelman RB, Orem W, Castranova V, Tatu CA, Belkin HE, Zheng B, Lerch HE, Maharaj SV, Bates AL (2002) Health impacts of coal and coal use: possible solutions. Int J Coal Geol 50:425–443
Flues M, Sato IM, Scapin MA, Cotrim MEB, Camargo IMC (2013) Toxic elements mobility in coal and ashes of Figueira coal power plant, Brazil. Fuel 103:430–436
Fujimori T, Tanino Y, Takaoka M (2011) Role of zinc in MSW fly ash during formation of chlorinated aromatics. Environ Sci Technol 45:7678–7684
Gupta N, Gedam VV, Moghe C, Labhasetwar P (2019) Comparative assessment of batch and column leaching studies for heavy metals release from coal fly ash bricks and clay bricks. Environ Technol Innov 16:100461
Gwenzi W, Mupatsi NM (2016) Evaluation of heavy metal leaching from coal ash-versus conventional concrete monoliths and debris. Waste Manag 49:114–123
Hagemeyer AN, Sears CG, Zierold KM (2019) Respiratory health in adults residing near a coal-burning power plant with coal ash storage facilities: a cross-sectional epidemiological Study. Int J Environ Res Public Health 16:3642
Hatungimana D, Taşköprü C, İçhedef M, Saç MM, Yazıcı Ş, Mardani-Aghabaglou A (2020) Determination of radon and natural radioactivity concentration in some building materials used in Izmir, Turkey. J Green Build 15:107–118
Höfer I, Kaltschmitt M, Beckendorff A (2017) Emissions from solid biofuel combustion, pollutant formation and control options. In: Meyers RA (ed) Encyclopedia of sustainability science and technology. Springer, New York, NY, pp 1–30
Huang K, Inoue K, Harada H, Kawakita H, Ohto K (2011) Leaching of heavy metals by citric acid from fly ash generated in municipal waste incineration plants. J Mater Cycles Waste Manage 13:118–126
Ishaq M, Jan FA, Shakirullah M, Ahmad I, Ali H (2013) Leaching capabilities of iodine monochloride (ICl) and diethylenetriamine pentaacetic acid (DPTA) at different concentrations and pH for some major and trace metals from coal and coal fly ash. Arab J Geosci 6:4109–4118
Iyer R (2002) The surface chemistry of leaching coal fly ash. J Hazard Mater 93:321–329

Izquierdo M, Querol X (2012) Leaching behaviour of elements from coal combustion fly ash: an overview. Int J Coal Geol 94:54–66

Jambhulkar HP, Shaikh SMS, Kumar MS (2018) Fly ash toxicity, emerging issues and possible implications for its exploitation in agriculture; Indian scenario: a review. Chemosphere 213:333–344

Jana A, Ghosh M, Sinha S, Jothiramajayam M, Nag A, Mukherjee A (2017) Hazard identification of coal fly ash leachate using a battery of cyto-genotoxic and biochemical tests in allium cepa. Archives Agron Soil Sci 63:1443–1453

Jankowski J, Ward CR, French D, Groves S (2006) Mobility of trace elements from selected Australian fly ashes and its potential impact on aquatic ecosystems. Fuel 85:243–256

Jones DR (1995) The leaching of major and trace elements from coal ash. In: Swaine DJ, Goodarzi F (eds) Environmental aspects of trace elements in coal. Springer

Jones KB, Ruppert LF, Swanson SM (2012) Leaching of elements from bottom ash, economizer fly ash, and fly ash from two coal-fired power plants. Int J Coal Geol 94:337–348

Junk G, Ford C (1980) Review of organic emissions from selected combustion processes. Ames Lab, IA (USA)

Karim MR, Khandaker MU, Asaduzzaman K, Razak HA, Yusoff SB (2019) Radiological risks assessment of building materials ingredients: palm oil clinker and fuel ash. Indoor Built Environ 28:479–491

Kashiwakura S, Ohno H, Matsubae-Yokoyama K, Kumagai Y, Kubo H, Nagasaka T (2010) Removal of arsenic in coal fly ash by acid washing process using dilute H2SO4 solvent. J Hazard Mater 181:419–425

Kim AG (2002) Physical and chemical characteristics of cub. proc. coal combustion byproducts and western coal mines, US department of interior, office of surface mining coal research center. Southern Illinois University at Carbondale, p 18

Košnář Z, Mercl F, Perná I, Tlustoš P (2016) Investigation of polycyclic aromatic hydrocarbon content in fly ash and bottom ash of biomass incineration plants in relation to the operating temperature and unburned carbon content. Sci Total Environ 563:53–61

Kosson DS, van der Sloot HA, Sanchez F, Garrabrants AC (2002) An integrated framework for evaluating leaching in waste management and utilization of secondary materials. Environ Eng Sci 19:159–204

Kristina MZ, Chisom O (2020) A review on fly ash from coal-fired power plants: chemical composition, regulations, and health evidence. Rev Environ Health, 20190039

Kukier U, Ishak CF, Sumner ME, Miller WP (2003) Composition and element solubility of magnetic and non-magnetic fly ash fractions. Environ Pollut 123:255–266

Leelarungroj K, Likitlersuang S, Chompoorat T, Janjaroen D (2018) Leaching mechanisms of heavy metals from fly ash stabilised soils. Waste Manage Res 36:616–623

Li Z, Li X, Ma H, Fan L (2017) Polychlorinated biphenyls in fly ashes collected from five coal-fired power plants in North China. In: IOP conference series: earth and environmental science. IOP Publishing, p 012054

Lindroos A-J, Ryhti K, Kaakkurivaara T, Uusitalo J, Helmisaari H-S (2019) Leaching of heavy metals and barium from forest roads reinforced with fly ash. Silva Fennica

Liu K, Xie W, Zhao Z-B, Pan W-P, Riley JT (2000) Investigation of polycyclic aromatic hydrocarbons in fly ash from fluidized bed combustion systems. Environ Sci Technol 34:2273–2279

Liu HH, Shih TS, Chen IJ, Chen HL (2008) Lipid peroxidation and oxidative status compared in workers at a bottom ash recovery plant and fly ash treatment plants. J Occup Health 50:492–497

Liu H, Banerji SK, Burkett WJ, Van Engelenhoven J (2009) Environmental properties of fly ash bricks. In: World of coal ash (WOCA) conference, pp 4–7

Lu X, Zhang X (2007) Radionuclide content and associated radiation hazards of building materials and by-products in Baoji, West China. Radiat Prot Dosimetry 128:471–476

Lu X, Chao S, Yang F (2014) Determination of natural radioactivity and associated radiation hazard in building materials used in Weinan, China. Radiat Phys Chem 99:62–67

Mani U, Prasad AK, Suresh Kumar V, Lal K, Kanojia RK, Chaudhari BP, Murthy RC (2007) Effect of fly ash inhalation on biochemical and histomorphological changes in rat liver. Ecotoxicol Environ Saf 68:126–133
Maresca A, Astrup TF (2014) Leaching from biomass combustion ash. In: 5th international symposium on energy from biomass and waste: Venice 2014. CISA Publisher
Masto RE, Sarkar E, George J, Jyoti K, Dutta P, Ram LC (2015) PAHs and potentially toxic elements in the fly ash and bed ash of biomass fired power plants. Fuel Process Technol 132:139–152
Mastral A, Callen M, Murillo R, Garcia T, Vinas M (1999) Influence on PAH emissions of the air flow in AFB coal combustion. Fuel 78:1553–1557
Meharg AA, Williams PN, Adomako E, Lawgali YY, Deacon C, Villada A, Cambell RC, Sun G, Zhu Y-G, Feldmann J (2009) Geographical variation in total and inorganic arsenic content of polished (white) rice. Environ Sci Technol 43:1612–1617
Mehra R, Kaur S, Prakash R (2019) Optimization of fly ash content in cement and assessment of radiological risk. Indoor Built Environ 29:286–292
Meij R (1994) Trace element behavior in coal-fired power plants. Fuel Process Technol 39:199–217
Mollon L, Norton G, Trakal L, Moreno-Jimenez E, Elouali F, Hough R, Beesley L (2016) Mobility and toxicity of heavy metal (loid) s arising from contaminated wood ash application to a pasture grassland soil. Environ Pollut 218:419–427
Mubeen I, Buekens A, Chen Z, Lu S, Yan J (2017) De novo formation of dioxins from milled model fly ash. Environ Sci Pollut Res 24:19031–19043
Mulder E (1996) Pre-treatment of MSWI fly ash for useful application. Waste Manag 16:181–184
Müller G, Grimmer G, Böhnke H (1977) Sedimentary record of heavy metals and polycyclic aromatic hydrocarbons in lake constance. Naturwissenschaften 64:427–431
Pan Y, Yang L, Zhou J, Liu J, Qian G, Ohtsuka N, Motegi M, Oh K, Hosono S (2013) Characteristics of dioxins content in fly ash from municipal solid waste incinerators in China. Chemosphere 92:765–771
Pandey VC, Abhilash PC, Singh N (2009) The Indian perspective of utilizing fly ash in phytoremediation, phytomanagement and biomass production. J Environ Manage 90:2943–2958
Pandey VC, Singh JS, Kumar A, Tewari DD (2010) Accumulation of heavy metals by chickpea grown in fly ash treated soil: effect on antioxidants. CLEAN Soil Air Water 38:1116–1123
Papaefthymiou H, Gouseti O (2008) Natural radioactivity and associated radiation hazards in building materials used in Peloponnese, Greece. Radiat Meas 43:1453–1457
Papastefanou C (2008) Radioactivity of coals and fly ashes. J Radioanal Nucl Chem 275(1):29–35
Patra K, Rautray TR, Nayak P (2012a) Analysis of grains grown on fly ash treated soils. Appl Radiat Isot 70:1797–1802
Patra KC, Rautray TR, Tripathy BB, Nayak P (2012b) Elemental analysis of coal and coal ASH by PIXE technique. Appl Radiat Isot 70:612–616
Praharaj T, Powell MA, Hart BR, Tripathy S (2002) Leachability of elements from sub-bituminous coal fly ash from India. Environ Int 27:609–615
Puettmann W, Schaefer RG (1990) Assessment of carbonization properties of coals by analysis of trapped hydrocarbons. Energy Fuels 4:339–346
Raclavska H, Raclavsky K, Matysek D (2009) Colour measurement as a proxy method for estimation of changes in phase and chemical composition of fly ash formed by combustion of coal. Fuel 88:2247–2254
Radke M, Willsch H, Teichmüller M (1990) Generation and distribution of aromatic hydrocarbons in coals of low rank. Org Geochem 15:539–563
Ram LC, Masto RE (2010) An appraisal of the potential use of fly ash for reclaiming coal mine spoil. J Environ Manage 91:603–617
Ram LC, Masto RE, Srivastava NK, George J, Selvi VA, Das TB, Pal SK, Maity S, Mohanty D (2014) Potentially toxic elements in lignite and its combustion residues from a power plant. Environ Monit Assess 187:4148

Rankers RH, Hohberg I (1991) Leaching tests for concrete containing fly ASH—Evaluation and mechanism. In: Goumans JJJM, van der Sloot HA, Aalbers TG (eds) Studies in environmental science. Elsevier, pp 275–282

Rantavaara A, Moring M (2000). Radioactivity of wood ash, Finland, p 43

Rautaray SK, Ghosh BC, Mittra BN (2003) Effect of fly ash, organic wastes and chemical fertilizers on yield, nutrient uptake, heavy metal content and residual fertility in a rice–mustard cropping sequence under acid lateritic soils. Biores Technol 90:275–283

Rodríguez-Barranco M, Lacasaña M, Aguilar-Garduño C, Alguacil J, Gil F, González-Alzaga B, Rojas-García A (2013) Association of arsenic, cadmium and manganese exposure with neurodevelopment and behavioural disorders in children: a systematic review and meta-analysis. Sci Total Environ 454–455:562–577

Roper AR, Stabin MG, Delapp RC, Kosson DS (2013) Analysis of naturally-occurring radionuclides in coal combustion fly ash, gypsum, and scrubber residue samples. Health Phys 104:264–269

Ross AB, Jones JM, Chaiklangmuang S, Pourkashanian M, Williams A, Kubica K, Andersson JT, Kerst M, Danihelka P, Bartle KD (2002) Measurement and prediction of the emission of pollutants from the combustion of coal and biomass in a fixed bed furnace. Fuel 81:571–582

Sarenbo S (2009) Wood ash dilemma-reduced quality due to poor combustion performance. Biomass and Bioenergy 33(9):1212–1220

Sahu SK, Bhangare RC, Ajmal PY, Sharma S, Pandit GG, Puranik VD (2009) Characterization and quantification of persistent organic pollutants in fly ash from coal fueled thermal power stations in India. Microchem J 92:92–96

Sanjuán MÁ, Suárez-Navarro JA, Argiz C, Mora P (2020) Assessment of natural radioactivity and radiation hazards owing to coal fly ash and natural pozzolan Portland cements. J Radioanal Nuclear Chem

Santhanam R, Balasundar S, Dhanalakshmi B, Meenakshisundaram V, Indira R, Baldev R (2005) Measurement of radioactivity content in flyash and its associated materials/buildings. In: International conference: fly ash India 2005, New Delhi, Fly Ash Utilization Programme (FAUP), TIFAC, DST, New Delhi

Sarode DB, Jadhav RN, Khatik VA, Ingle ST, Attarde SB (2010) Extraction and leaching of heavy metals from thermal power plant fly ash and its admixtures. Polish J Environ Studies 19:1325–1330

Sears CG, Zierold KM (2017) Health of children living near coal ash. Glob Pediatr Health 4:2333794X17720330–12333794X17720330

Singh A, Sharma RK, Agrawal SB (2008) Effects of fly ash incorporation on heavy metal accumulation, growth and yield responses of Beta vulgaris plants. Biores Technol 99:7200–7207

Singh S, Ram LC, Masto RE, Verma SK (2011) A comparative evaluation of minerals and trace elements in the ashes from lignite, coal refuse, and biomass fired power plants. Int J Coal Geol 87:112–120

Singh PK, Tripathi P, Dwivedi S, Awasthi S, Shri M, Chakrabarty D, Tripathi RD (2016) Fly-ash augmented soil enhances heavy metal accumulation and phytotoxicity in rice (Oryza sativa L); a concern for fly-ash amendments in agriculture sector. Plant Growth Regul 78:21–30

Steenari BM, Karlsson LG, Lindqvist O (1999) Evaluation of the leaching characteristics of wood ash and the influence of ash agglomeration. Biomass Bioenerg 16:119–136

Stout SA, Emsbo-Mattingly SD (2008) Concentration and character of PAHs and other hydrocarbons in coals of varying rank–implications for environmental studies of soils and sediments containing particulate coal. Org Geochem 39:801–819

Temuujin J, Surenjav E, Ruescher CH, Vahlbruch J (2019) Processing and uses of fly ash addressing radioactivity (critical review). Chemosphere 216:866–882

Till M, Behnisch P, Hagenmaier H, Bock KW, Schrenk D (1997) Dioxinlike components in incinerator fly ash: a comparison between chemical analysis data and results from a cell culture bioassay. Environ Health Perspect 105:1326–1332

Tiwari MK, Bajpai S, Dewangan UK, Tamrakar RK (2015) Suitability of leaching test methods for fly ash and slag: a review. J Radiat Res Appl Sci 8:523–537

Tsiridis V, Petala M, Samaras P, Kungolos A, Sakellaropoulos GP (2012) Environmental hazard assessment of coal fly ashes using leaching and ecotoxicity tests. Ecotoxicol Environ Saf 84:212–220
Ugurlu A (2004) Leaching characteristics of fly ash. Environ Geol 46:890–895
Ullah H, Liu G, Yousaf B, Ali MU, Abbas Q, Zhou C (2017) Combustion characteristics and retention-emission of selenium during co-firing of torrefied biomass and its blends with high ash coal. Biores Technol 245:73–80
USEPA (2010) US environmental protection agency. hazardous and solid waste management system; identification and listing of special wastes; disposal of coal combustion residuals from electric utilities; proposed rule. Fed Reg 75:35128–35264
Van Gerven T, Van Keer E, Arickx S, Jaspers M, Wauters G, Vandecasteele C (2005) Carbonation of MSWI-bottom ash to decrease heavy metal leaching, in view of recycling. Waste Manage 25:291–300
Verma SK, Masto RE, Gautam S, Choudhury DP, Ram LC, Maiti SK, Maity S (2015) Investigations on PAHs and trace elements in coal and its combustion residues from a power plant. Fuel 162:138–147
Wang T, Wang J, Tang Y, Shi H, Ladwig K (2009) Leaching characteristics of arsenic and selenium from coal fly ash: role of calcium. Energy Fuels 23:2959–2966
Wang N, Sun X, Zhao Q, Yang Y, Wang P (2020) Leachability and adverse effects of coal fly ash: a review. J Hazard Mater 396:122725
Warren CJ, Dudas MJ (1984) Weathering processes in relation to leachate properties of alkaline fly ash. J Environ Qual 13:530–538
White PJ, Broadley MR (2009) Biofortification of crops with seven mineral elements often lacking in human diets–iron, zinc, copper, calcium, magnesium, selenium and iodine. New Phytol 182:49–84
Williams PT, Bartle KD, Andrews GE (1986) The relation between polycyclic aromatic compounds in diesel fuels and exhaust particulates. Fuel 65:1150–1158
Xue J, Liu G, Zhang H, Niu Z (2006) Research on the impact of PAHs extracted from raw coal in different times. Res Environ Sci 19:107–112
Yan R, Gauthier D, Flamant G (2001) Partitioning of trace elements in the flue gas from coal combustion. Combust Flame 125:942–954
You X (2008) Polycyclic aromatic hydrocarbon (PAH) emission from co-firing municipal solid waste (MSW) and coal in a fluidized bed incinerator. Waste Manage 28:1543–1551
Yu C-L, Deng Q, Jian S, Li J, Dzantor EK, Hui D (2019) Effects of fly ash application on plant biomass and element accumulations: a meta-analysis. Environ Pollut 250:137–142
Zevenbergen C, Wood TV, Bradley J, Van der Broeck P, Orbons A, Van Reeuwijk L (1994) Morphological and chemical properties of MSWI bottom ash with respect to the glassy constituents. Hazard Waste Hazard Mater 11:371–383
Zevenbergen C, Bradley JP, van Reeuwijk LP, Shyam AK, Hjelmar O, Comans RNJ (1999) Clay formation and metal fixation during weathering of coal fly ash. Environ Sci Technol 33:3405–3409
Zha X, Ning J, Saafi M, Dong L, Dassekpo J-BM, Ye J (2019) Effect of supercritical carbonation on the strength and heavy metal retention of cement-solidified fly ash. Cem Concr Res 120:36–45
Zhao Z-B, Liu K, Xie W, Pan W-P, Riley JT (2000) Soluble polycyclic aromatic hydrocarbons in raw coals. J Hazard Mater 73:77–85
Zielinski RA, Finkelman RB (1997) Radioactive elements in coal and fly ash: abundance, forms, and environmental significance. US Geological Survey
Zierold KM, Hagemeyer AN, Sears CG (2020) Health symptoms among adults living near a coal-burning power plant. Arch Environ Occup Health 75:289–296

Chapter 5
Conclusions

Abstract Coal and biomass will continue to remain as the major source for energy mainly due to their abundance, availability, and affordability. More and more ash will be generated in the coming years. Most of these ashes have amicable properties that support their use in large volume and value added sectors. However, few ashes have high amount of potential toxic substances that needs to be monitored before its disposal and utilisation.

Coal is, and will continue to remain, the backbone of energy economics for India and several other countries, for the coming decades, mainly due to its abundance, availability, and affordability. However, because of its high carbon foot-prints and global warming potentials, many countries have advocated the use of biomass feedstocks for energy generation. Though these two fuel-feedstocks originated from plants, the environmental conditions that prevailed during their genesis had significantly affected the nature of the mineral matter in the feedstock and accordingly the ash generated after combustion has distinct characteristics. Fly ash and bottom ash, as the respective names indicate, are the dominant contrasting products formed from coal combustion. While combustion of coal is known for its deleterious impact on the atmosphere, biomass combustion which has also gained much attention in recent years is believed to have less negative impact on the atmosphere due to its CO_2 neutral-conversion.

In the past few decades, the subject of coal and biomass combustion products/by-products (such as fly ash and bottom ash) has emerged as a point of discussion due to the huge generation across the challenges faced by the power plants for utilization and disposal of the ash. The relation between the ash properties and their potential utilization is a complex subject and warrants a thorough understanding of the genesis, physical and chemical properties, potential environmental hazards, and their abatement strategies. In this monograph, on a step-by-step basis we have documented the importance of the subject, the characterization techniques, the utilization potential of coal and biomass ash, and the environmental concerns associated with it. Comparative evaluation of coal and biomass ash in terms of their properties to potential utilization is the key novel feature of this work.

A. K. Singh et al., *Ash from Coal and Biomass Combustion*,
https://doi.org/10.1007/978-3-030-56981-5_5

In terms of characterization, different ash properties, viz., particle size distribution, porosity, major oxide composition, unburnt carbon (LOI), and micro-morphology play a critical role in developing the ash utilization plan. Coal ash is primarily composed of oxides of Si, Al, Fe, and Ca, whereas Si, Ca, K, Na are the dominant constituents of biomass ash. Being an organic-sedimentary rock, coal inherently consists of different types of mineral matter with Si and Al as dominant inorganic phases. The mineral content in coal can vary largely depending upon their mode of origin and deposition, whereas biomass has low ash content due to the phyto origin of the inorganic constituents. As the character of the coal and biomass controls the nature of the ash produced, thorough understanding of their properties is an essential must, and to predict their potential environmental impacts and their utilization potential. For instance, ash generated from biomass-based power plant will have potential for extraction of potassium, while ash from pet coke combustion has the potential to extract V and Ni.

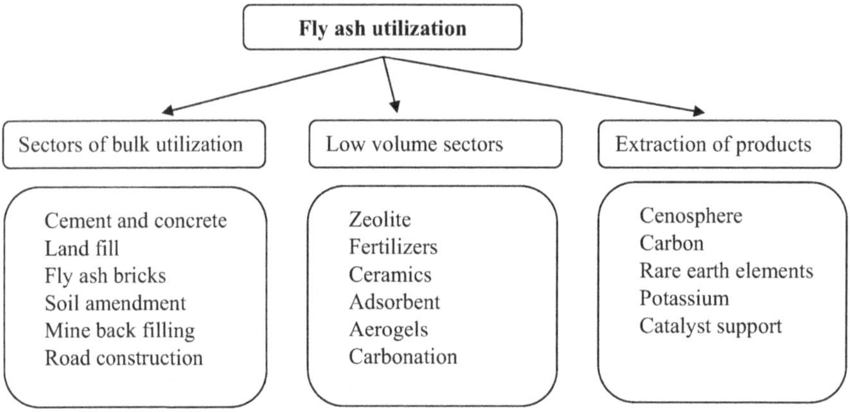

Fly ashes are heterogeneous, and their property varies with the nature of the coal/biomass; and nature of combustion (especially the boiler temperature), and ash handling and collection system. The chemical composition and physical properties of the ash decide its potential utilization. The desired properties for different utilization options are reviewed in the monograph. Modification of ash properties to suit the requirements of specific utilization is also discussed. The favorable silt size particles of ash along with the presence of plant nutrients in coal fly ash favors its use in soil reclamation, whereas presence of potentially toxic elements in bio-available forms limits its utilization for agriculture purposes. Presence of S, Al, and Ca oxides in fly ash imparts pozzolanic properties suitable for its use in cement industry, however, the pozzolanic properties of biomass ash are relatively less and make it less suitable in cement industry. The pozzolanic property of coal ash also supports soil aggregation, and thereby amends soil bulk-density, porosity, and water-retention capability. Predominance of silt-sized particles in ash modifies soil texture and infiltration rate. Occurrence of significant amount of plant nutrients in the ash improves the nutrient pool of the soil, and these in turn aids in the proliferation of soil organisms and the overall soil health. Pozzolanic properties coupled with plasticity of coal ash favors its suitability for brick making, whereas presence of alkalis is the major limitation for the use of biomass ash. Leaching of potassium and alkalis from biomass ash makes its suitable for brick making. The potential sectors of ash utilization including agriculture and forestry; cement and concrete; bricks, tiles, blocks, and lightweight aggregates; embankment and road construction; mine filling; reclamation of low-lying area; and extraction of value-added materials are detailed in the monograph. Inclusion of some of the niche areas of ash utilization adds value to the monograph. Specialized areas like use of fly ash as a catalyst; waste water treatment; adsorption of mercury and toxic elements; potassium fertilizer; nitride fertilizer; and for carbon sequestration have been discussed.

One of the major apprehensions for fly ash utilization is the onsite and offsite potential environmental hazards. The content of potentially toxic elements, PAHs, and their potential environmental health hazards are critical. Knowledge on the behavior of potentially toxic elements during combustion in power plants especially the fate of volatile and semi-volatile elements and their partitioning between fly ash and bottom ash is essential. It is this partitioning, volatilization-condensation, that exclusively causes concentration of arsenic from pyrites in coal, within fly ash component, than so in the bottom ash component. Knowledge on chemical association of potentially toxic elements and their fractionation into water soluble, exchangeable, carbonate bound, metal oxide bound, organic fraction and silicates bound forms in coal and biomass ashes generated from diverse feedstocks are essential. Cases of transfer of toxic substances into the real soil, water, and air media, as well as in the food product grown from fly ash added soils, and all products made from ash, if any, are also extremely critical. Consequently, abatement of fly ash toxicity using physical, chemical, and biological means and the need for periodical monitoring have also gained research interest in recent years. The polycyclic aromatic hydrocarbon (PAH) compounds present within coal are of environmental and human health concerns, and may concentrate in the ash fractions or the gaseous fractions produced from coal combustion. PAH concentrations in coal fly and bottom ash correlates with combustion conditions and the character of the coal being used in the power plant. Beyond coal ashes, unburnt coal as a source of PAH leading to contamination of associated sediments/soils has also been established in the recent past. Biomass ashes produced from biomass combustion are also marked by the presence of PAHs. Pyrolysis and pyro-synthesis are two processes that are accepted widely, to cause formation of PAHs in coal and biomass ashes.

Some of the practical bottlenecks for ash utilization are the heterogeneity of ash properties, general environmental apprehension, cost of transportation and contamination during transport, lack of policies and regulatory directives, etc. Global ash generation will continue to increase as the energy requirement is increasing. However, it is envisaged that the contribution of biomass feedstock will increase and there will be many new power plants with co-firing of coal-biomass feed stocks. The future scope of ash research would be primarily on the ash generated from co-combustion of coal and biomass. Overall, this monograph is a unique piece of work to get comprehensive information on coal and biomass ash simultaneously and help the researchers and power plant managers to pitch the way forward for handling and sustainable utilization of coal and biomass ashes.

The ashes from coal and biomass combustion, in recent years, have emerged as a valued commodity, due to their heterogeneous and diverse properties, and have found multi-faceted utilization in different sectors, viz., building materials, agriculture and forestry, source of value-added products. India, by the year 2030, is estimated to be having excess ash, beyond its utilization in different sectors. Consequently, in the coming years the excess ash produced in the country can act as valued 'trade' commodities for other countries which are in dearth of ash for its utilization in different sectors. Middle East, Australia, Africa and few others are believed to be the key markets for Indian ash export. However, international trade of ash would likely to face a lot of barriers, viz., the expenditure toward inland-shipping, expenditure at ports, likely idea of acceptance of supply for only large consignments, shipping consent, and so on. In India, with push in infrastructures and funding in the coming years, the ash trade drive can potentially grow.

The manufacturer's authorised representative in the EU is Springer Nature Customer Service Centre GmbH, Europaplatz 3, 69115 Heidelberg, Germany. If you have any concerns regarding our products, please contact ProductSafety@springernature.com

Printed and bound by CPI Group (UK) Ltd, Croydon, CR0 4YY
30/09/2026
02225574-0001